Climate Change

Thomas Brewer

Climate Change

An Interdisciplinary Introduction

Springer

Thomas Brewer
Emeritus Faculty, School of Business
Georgetown University
Washington, DC, USA

ISBN 978-3-031-42905-7 ISBN 978-3-031-42906-4 (eBook)
https://doi.org/10.1007/978-3-031-42906-4

Legal and Financial Statements: As the author, I have produced the book as an independent scholar. I have no financial conflicts of interest. Nothing in the book should be interpreted as offering legal or financial advice.

This Springer imprint is published by the registered company Springer Nature Switzerland AG
The registered company address is: Gewerbestrasse 11, 6330 Cham, Switzerland

Paper in this product is recyclable.

For all the world's children

Preface

As the climate change agendas for governments, businesses, and civil societies have expanded dramatically, it has become increasingly urgent to address a wide range of issues. An earlier focus on gradual, long-term mitigation of carbon dioxide emissions is no longer sufficient, though it is still surely necessary. In addition, there are now complementary goals concerning short-lived climate pollutants (SLCPs) such as methane and black carbon that are receiving more intensive attention. Public opinion in some countries has also shifted to favor more serious short-term actions by governments and businesses to mitigate emissions and adapt to their effects. Local adaptation issues have also become more salient because of an increase in the frequency and intensity of extreme weather events that are being attributed to climate change.

The climate change agenda has shifted to include more focus on *actions*—actions by governments, by businesses, by other kinds of organizations, and by individual citizens. The book is therefore not only about analysis of the issues, but also about actions that have been taken and further actions that can be taken.

The book thus includes a broad range of topics, and it approaches them from an inter-disciplinary perspective, which offers breadth of coverage. The disciplinary perspectives include climate science, economics, political science, history, international law, ethics, and industry-specific technologies. The need for holistic inter-disciplinary analysis is a recurrent theme. The approach is therefore consistent with the IPCC Synthesis report (2023: 3; italics added) that "recognizes the interdependence of climate, ecosystems and biodiversity, and human societies; *the value of diverse forms of knowledge*; and the close linkages between climate change adaptation, mitigation, ecosystem health, human well-being and sustainable development, and reflects the increasing diversity of actors involved in climate action."

My research for the book has relied extensively on refereed journal articles, official reports by inter-governmental international organizations and national governments, as well as think tanks, and other non-profit non-governmental organizations. I have also used many periodical news organizations for keeping up with recent developments.

Chapter 1 introduces the wide variety of issues that involve a mix of problems and solutions. Chapter 2 presents key issues, concepts, and data from climate science. Chapter 3 analyzes the emissions, technologies, economies, and policies of major emitters—including China, India, the USA, Germany, and other countries in the European Union. It also focuses on populations in all parts of the world where people are particularly vulnerable to the effects of climate change and where issues about adaptation policies are salient; that chapter includes ethical issues as well as international economic and political issues. Chapter 4 applies historical and legal analysis to international agreements concerning mitigation and adaptation.

Chapters 5–11 combine corporate, technological, and governmental policy topics for economic sectors. The section begins with chap. 5 on fossil fuels. Successive chapters focus on the electric power, transportation, and industry sectors—which together account for more than half of the world's greenhouse gas and particulate emissions. Chapters that focus on industries, buildings, food, and forestry are mostly concerned with the production of diverse kinds of tangible goods, while banking and insurance are primarily service-oriented but also involve many issues of direct relevance to all other sectors.

Chapter 12 discusses scenarios of the future based on projections from climate science studies. Although the chapter is inevitably speculative about the future, it is also grounded in empirical analyses of the climate changes and the responses by governments, businesses, and civil society to date. Tipping points receive special attention, particularly in regard to the consequences of ice melting in the Arctic and Antarctic regions.

The book is eclectic in its coverage of countries and regions. It includes data, examples, and discussions about many countries in all regions of the world. Countries with relatively large shares of global emissions receive special attention, as do countries that are especially vulnerable to the impacts of climate change. In addition, regardless of their size, actors that are leaders in key technological or policy developments concerning mitigation of emissions or adaptation to their impacts are also subject to special attention.

There are two short lists at the end of each chapter. One is a list of "questions to ponder" that can be used by individual readers and/or groups to focus attention on key issues. Some are scientific or engineering, while others are political, economic, historical, or ethical. The other list is "resources for keeping up with developments," which can be especially helpful for further reading or research.

The book, therefore, also offers depth of coverage of topics that are especially salient, or complex, or controversial, or need more attention than popular discourse is granting them. These topics are addressed in boxes within the bodies of the chapters and annexes to the chapters. Research on the book ended on July 7, 2023.

Finally, a short note about people's reactions to climate change issues. Extreme weather events have become more frequent, more intensive, and more destructive in the early 2020s, and climate science projections of the future have become more alarming. As a result, there has understandably been an increase in concern among climate experts, journalists, and mental health professionals about the effects of climate change.

Although this textbook does not focus on these issues, they are part of the contemporary context of climate change topics. It is thus appropriate to make explicit my views on the central issues. I reject the following: (1) denial of the existence of climate change; (2) dismissal of projections of the future because of uncertainties; and (3) refusal to think about catastrophic scenarios because they seem too frightening.

These responses are inconsistent with the following: (1) The denial response has been discredited by thousands of experts and massive empirical research over many years in countries all over the world. (2) As projections about the future are being periodically updated by additional research, many problems are projected to be more serious and likely to happen sooner than previously thought. (3) Although there are obviously many imminent problems related to climate change—some of them even "existential"—it is also true that there are many solutions to them. In fact, there are *mitigation and adaptation actions* that can be taken—and are being taken—by governments, businesses, NGOs, and individuals that can reduce the problems. There are indeed many solutions as well as many problems.

Washington, DC, USA Thomas Brewer

Reference

IPCC (2023). Synthesis Report of the IPCC Sixth Assessment Report (AR6), Summary for Policymakers. https://report.ipcc.ch/ar6syr/pdf/IPCC_AR6_SYR_SPM.pdf Accessed 21 March 2023.

Acknowledgements

Writing an interdisciplinary text about climate change poses distinctive and daunting professional challenges. I have been fortunate to have had much help from experts in universities, governments, international organizations, non-governmental organizations, corporations, and students.

In 1999, while reading the annual reports of several environmental organizations, I became more aware of the implications of climate change, and I began to wonder if I could do useful research on the topic. During the 1999–2000 academic year, when I was living in Denmark, it became clear that there was already a consensus that climate change was a serious problem that was going to get worse; the question was what to do about it.

During that time, as a participant in a World Bank conference in Kenya on environmental issues in developing countries, I became aware of plans for a short summer course on climate change in 2000 at Harvard. The course was organized by Rob Stavins, with the assistance of Kelly Sims Gallagher, and it was an enriching experience that reinforced my determination to focus my research on climate change issues and to do it in an interdisciplinary way.

After I returned to the Georgetown Business School, I was able to introduce an MBA course on "Energy and Climate Change Issues in Business." I also got to know Vicky Arroyo, who was developing the Georgetown Climate Center at the Law School—a center that has become a significant resource for government officials and others whose professional responsibilities include climate change issues at the local, state, and federal government levels in the USA. I also had colleagues in the Georgetown business school who were supportive over the years—including Stanley Nollen, Dennis Quinn, Gene Salorio, and Dean Robert Parker.

I have been fortunate to be a visiting scholar or an instructor at many universities, where there were opportunities to make presentations to faculty and student groups. They included David King at Oxford, Michael Mehling at MIT, and Andreas Falke at Nuremburg. I was also a visiting international scholar in France at Strasbourg and St. Germain.

Via my affiliation with the International Center on Trade and Sustainable Development (ICTSD) in Geneva, I was fortunate to learn a lot from Ricardo Menendez-Ortiz, Andrew Crosby, Ingrid Jigou, and Mahesh Sugathan. I also learned much from Christian Egenhofer and his colleagues while writing reports and attending conferences at the Centre for European Policy Studies (CEPS) in Brussels. As a participant in a series of workshops on international maritime issues organized by the Brian Comer and Dan Rutherford at the International Council on Clean Transportation (ICCT), based in Washington, DC, enabled me to become conversant with the climate change issues in that industry. Another excellent opportunity to pursue my international martime shipping interests was being a contributor to a volume on short-lived climate pollutants edited by Yulia Yamineva at the Centre for Climate Change, Energy and Environmental Law at the University of Eastern Finland.

I have learned much from many co-authors, including Harro van Assalt, Michael Mehling, Michael Grubb, and Henry Derwent.

I am indebted to many people who have read and commented on parts of chapters or in some cases entire chapters. They were all consistently helpful, whether as topical experts or general readers. They are: Marianne Asmussen, Jennifer Brewer, Sandra Brewer, Andrew Crosby, Joe Dahmen, Susan Emanuel, Janice Harmeier, Lise Kagenow, and Fabienne Spier.

At Springer Nature publishing company, Editor Anthony Doyle and Project Coordinator Manju Ramanathan were both consistently helpful from beginning to end and at several points along the way. Thanks and thanks!.

Contents

Part I
Introduction

Overview of the Issues

1

> The deadly impacts of climate change are here and now.
>
> The good news is that we know what to do and we have the financial and technological tools to get the job done.
>
> UN Secretary General Antonio Guterres at COP 27, November 7, 2022
>
> (Guterres 2022)

1.1 Introduction

A recurrent theme of this book is that the topic “climate change” is not simply one problem with one solution; rather “climate change” is a short-hand expression that includes many problems and many solutions. This brief introductory chapter first anticipates the many problems and solutions that are addressed in the individual chapters.

The inter-disciplinary approach of the book is useful for analyzing both problems and solutions. Climate science topics are thus combined with economic, political, legal, ethical, historical and technological perspectives to promote a broad understanding of the challenges posed by climate change. Of course, an understanding of the concepts, logic and evidence of climate science is an essential foundation for an understanding of the many specific problems and solutions.

T. Brewer, *Climate Change*, https://doi.org/10.1007/978-3-031-42906-4_1

1.2 Climate Science Basics

Climate change forcing agents include many kinds of gasses as well as the particulate matter black carbon. Carbon dioxide is the principal long-term greenhouse gas with an average atmospheric lifetime of a century; nitrous oxide gas is also an important long-term forcing agent. In the shorter-term of two decades, methane is more than eighty times more potent per tonne than carbon dioxide, and black carbon particulate matter is more than a thousand times more potent per tonne than carbon dioxide. Methane's average atmospheric lifetime is about 12 years, and black carbon's is about one week (Chap. 2).

Climate change problems will continue to get worse because the concentration level of carbon dioxide will continue to increase until a *net-zero annual rate* of carbon dioxide emissions is reached—that is, when global emissions are fully absorbed by oceans, forests and other sinks. A widely accepted target is to achieve net zero by 2050 (Chap. 2).

There are many and wide-ranging *direct effects of climate change including extreme weather events*—not only heat waves, but also droughts, wildfires, floods, and the most severe tropical cyclones (which are also known as hurricanes or typhoons). Such effects are already happening in all regions of the world. In coastal areas, higher *sea levels* are causing increased damage from storms and floods (Chaps. 2, 3).

1.3 Socio-Economic Impacts, Mitigation Actions, Adaptation Actions, and Ethics

Impacts on societies include: lost incomes; damaged infrastructures; mass migrations internationally and within countries; and increased health problems, including mental health problems, as well as human deaths (Chaps. 2, 3).

Short-term socio-economic impacts are not distributed equally among groups. The impacts include floods among people living near coasts or rivers, droughts among people living near deserts, and wildfires among people living near forests. There are also increased disparities in incomes among countries and among groups of people within countries; poor countries and poor groups within countries are more vulnerable to the socio-economic impacts than are rich countries and rich groups within countries. *However, over time the socio-economic impacts spread to entire countries and regions.* Issues between the group of wealthy countries that are also major emitters, on the one hand, and poor countries that suffer from the emissions, on the other hand, not only pose international political and economic conflicts but also ethical issues (Chap. 3).

Ethical issues are inherent in the economic disparities between emitters and those who suffer from the effects of their emissions. The ethical issues can be analyzed empirically; the analyses can be formulated, for instance, in terms of public health indicators, such as numbers of human lives lost or saved or total numbers of person-years of longevity lost in relation to life expectancy. Economic impacts can also be calculated in terms of the monetary benefit–cost analyses of action or inaction scenarios—and the distributions of the benefits and costs among groups of people (Chap. 3).

There are numerous examples of climate action leaders among governments, businesses, and civil society organizations (Chaps. 3–11).

Government policies at all levels—and socially responsible individual and organizational actions of many types—can mitigate emissions of climate change forcing agents and/or increase adaptation measures to reduce the emissions' human and socio-economic impacts. The government policies include regulations, taxes, subsidies and the creation of emission trading systems (Chaps. 3–12).

International conferences, agreements and institutions are important components of global and regional arrangements that facilitate the adoption of mitigation and adaptation measures (Chap. 4).

As government regulations change and consumer behaviors change, corporations find new opportunities for self-interested actions that can reduce their carbon footprints. In addition, changes in public knowledge, opinions and actions about climate change issues can change the self-interested calculations and actions of governments and corporations (Chaps. 5–11).

Some climate actions are more *economically cost-effective* and/or *politically more feasible* than others (Chaps. 5–11).

1.4 Economic Sectors and Industries

There are significant variations among *economic sectors* and among *industries* within sectors in the mix of problems and solutions, but there are also many cross-sector and cross-industry problems and solutions (Chaps. 5–11).

Some industries have already adopted significant technological changes and promulgated mandatory local and/or national and/or international regulations to incentivize and enforce the adoption—for instance in the motor vehicle and maritime shipping industries (Chaps. 5–11).

Actions that change business strategies and practices are pertinent for individual corporations, business associations, and managerial functions (Chaps. 5–11).

Some solutions involve *complex combinations* of technological change, government policies, business practices, and market-determined prices (Chaps. 5–11).

The changing economic and political conditions—and expectations about further future changes—create incentives for *technological innovations* in many industries, including not only in the carbon-intensive fossil fuel, electric power and transportation sectors, but other sectors as well (Chaps. 5–11).

1.5 The Future

Future climate conditions, compared with past and present conditions, are inherently more speculative and vulnerable to the distortions of a person's own preferences. Therefore, future scenarios that are developed from *evidence-based models of climate science are especially important* (Chap. 12).

Tipping points that could mark irreversible catastrophic changes are attracting more attention as emission trends and projections become more threatening (Chap. 12).

Proposals for geo-engineering systems to reduce the entry of climate change forcing agents into the earth's atmosphere are also attracting more interest (Chap. 12).

Shocks from economic systems, political systems , wars, technological innovations, health pandemics and other sources can significantly alter climate change patterns and trends. While some shocks exacerbate climate change, others have ameliorative effects (Chap. 12).

The political economy of the distribution of the benefits and costs over time and among groups of people often pose significant constraints on governments, businesses, international organizations, and public groups as they contemplate actions—and inaction (Chap. 12).

1.6 Analytic Challenges

The complexities of climate change issues are abundantly evident in the periodic reports of the Intergovernmental Panel on Climate Change (IPCC).

In its sixth Assessment Report in 2022, the IPCC summarized climate change as "unprecedented in its *scope* (sectors, actors and countries), *depth* (major transformations) and *time scales* (over generations). As such, it creates unique challenges for analysis" (IPCC 2022c; italics added).

An example of the complexities of a multi-dimensional, inter-disciplinary issue is an assessment of the "overall feasibility of limiting warming to 1.5 °C, and the feasibility of adaptation and mitigation options compatible with a 1.5 °C warmer world." Such an analysis involves six "dimensions" (IPCC 2018: 71; italics added):

- *Geophysical*: What global emission pathways could be consistent with conditions of a 1.5 °C warmer world? What are the physical potentials for adaptation?
- *Environmental-ecological*: What are the ecosystem services and resources, including geological storage capacity and related rate of needed land-use change, available to promote transformations, and to what extent are they compatible with enhanced resilience?
- *Technological*: What technologies are available to support transformation?
- *Economic*: What economic conditions could support transformation?

- *Socio-cultural*: What conditions could support transformations in behavior and lifestyles? To what extent are the transformations socially acceptable and consistent with equity?
- *Institutional:* What institutional conditions are in place to support transformations, including multi-level governance, institutional capacity, and political support?

These and other perspectives are applied to a wide range of questions in the following chapters in the book. Each chapter includes a list of questions for further consideration and a list of resources for keeping track of new developments.

References

Intergovernmental Panel on Climate Change (IPCC). (2018). Cross-Chapter Box 3, Framing Feasibility: Key Concepts and Conditions for Limiting Global Temperature Increases to 1.5 °C. Global Warming of 1.5 °C. *An IPCC Special Report*. Contributing Authors: Solecki, W., Cartwright, A., Cramer, W., Ford, J., Jiang, K., Pereira, J., Rogelj, J., Steg, L., & Waisman. H. Cambridge University Press, Cambridge, UK and New York, NY, USA. https://doi.org/10.1017/9781009157926.003

Intergovernmental Panel on Climate Change (IPCC). (2022c). Introduction and Framing. Climate Change 2022: Mitigation of Climate Change. Contribution of Working Group III to the Sixth Assessment Report of the Intergovernmental Panel on Climate Change. By M. Grubb, C. Okereke, J. Arima, V. Bosetti, Y. Chen, J. Edmonds, S. Gupta, A. Köberle, S. Kverndokk, A. Malik, & L. Sulistiawati. Cambridge University Press, Cambridge, UK and New York, NY, USA. https://doi.org/10.1017/9781009157926.003

Part II
Core Issues

Emissions: Types, Effects and Sources 2

> Human influence has warmed the climate at a rate that is unprecedented in at least the last 2000 years. … Human-induced climate change is already affecting many weather and climate extremes in every region across the globe.
>
> Intergovernmental Panel on Climate Change (IPCC, 2021a)

2.1 Introduction

This chapter introduces the types of greenhouse gas and particulate emissions, and it highlights their profiles as forcing agents of climate change. The chapter introduces key concepts and information about the emissions that are causing climate change, the effects of the emissions, and their sources. The chapter also explains the statistical indicators and mathematical models that are used to track and project emissions.

The chapter begins with a series of charts that are essential to understanding key emission patterns and trends. The sources of emissions are documented, with an introduction to the major emitting countries that are sources of the emissions. The effects of increasing emissions are discussed, with a focus on extreme weather events of recent years, as well as the long-term trends in sea-level rise, hurricanes, floods, droughts, crop failures, and wildfires. Evidence of the increasing impacts on human health and economies is also noted.

The annexes provide additional analyses of topics of special interest: attribution of the causes of extreme weather events; warming in the Arctic; and Models, Pathways and Scenarios.

T. Brewer, *Climate Change*, https://doi.org/10.1007/978-3-031-42906-4_2

2.2 Carbon Dioxide Emission Rates and Concentration Levels

Figure 2.1 depicts the global *yearly emission rates* and the *cumulative concentration levels* of carbon dioxide (CO_2) in the atmosphere from 1750 to 2019. The year 2019 was the last "normal" year prior to the covid pandemic of 2020–2022, which temporarily disrupted the long-term trends. The long-term increases, especially since the early 1960s, are among the core problems of climate change. The net annual increases in the emissions rates over time, of course, account for the increases in the concentration levels. The emissions are annual flows, while the concentration levels are the accumulated amounts in the atmosphere at the end of the year, or sometimes the average for the year. It is the increasing global concentration level that is a principal direct driver of global warming. The accumulated concentration level, however, could be reduced by continual net-zero annual emissions—which has become a widely recognized target.

Because the average atmospheric lifetime of CO_2 is about 100 years, approximately half of the molecules emitted during a given year will still be in the atmosphere a century later. In the meantime, subsequent CO_2 emissions over the century will contribute to the concentration levels. *Thus, reducing CO_2 emissions in one year does not necessarily reduce the concentration level.* This happened in 2020, when the covid pandemic reduced economic activity, and the CO_2 *emissions rate decreased* by 8.8% during the first half of 2020 (Liu et al. 2020); however, the CO_2 *concentration level continued to increase* during the same period by about 1%.

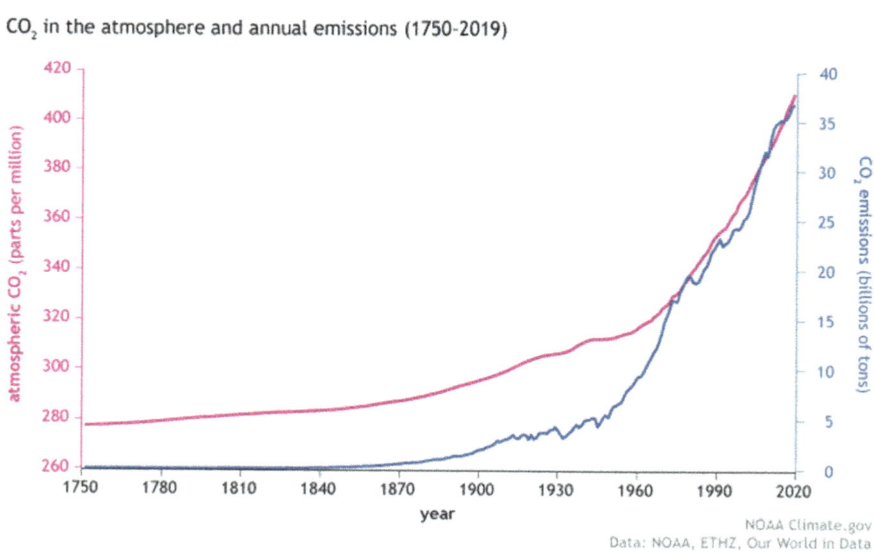

Fig. 2.1 Carbon dioxide global emissions rates and concentration levels over time. *Source* US NOAA (2021b)

A year later the concentration level was even higher at 419 parts per million (ppm), as recorded by the UK Met Office (2021) and the Mauna Loa observatory in Hawaii (US NOAA, 2021). The 419 ppm in mid-2021 was the highest in the US since 1958, when record keeping began for the US levels, and the highest since 1750 according to UK reconstructed estimates. The peak in 2022 was even higher at 421, which was the highest in millions of years (US NOAA, 2022a).

Two features of the units of measurement in Fig. 2.1 should be noted. First, the scales are different: parts per million (ppm) for concentration levels on the left vertical axis for the thin, light line versus billions of "tons" of emissions on the right vertical axis for the thick, dark line. Second, CO_2 emissions in 2019 were 36.44 billion *metric tonnes* or the equivalent of 40.17 billion non-metric tons. Sometimes "ton" is used as a spelling alternative to "tonne," as it is in Fig. 2.1. However, the term "ton" can also be used to mean a non-metric ton, which is 0.907 as much as a metric "tonne." This book typically uses metric data identified as "tonne" because that is virtually universal usage among scientists. (There are only a few "non-metric countries"—one being the US, whose climate scientists of course use metric measures in their professional work.).

2.3 Temperature Changes

In Fig. 2.2, the changes in CO_2 emissions and the changes in temperatures over hundreds of thousands of years are based on ice core samples from Antarctica (US NASA, 2021; US NOAA, 2020, 2021). The close correspondence between the CO_2 concentration levels and temperature changes is evident (1 °C = 1.8 °F). Of course, such close correlation does not necessarily by itself reflect a causal relationship. However, the existence of a causal connection in this case can be explained by atmospheric chemistry and physics: The CO_2 and other greenhouse gasses in the atmosphere "work mostly by preventing convection currents from carrying away heat absorbed from sunlight, not by preventing heat from radiating away from a surface"—as the concept of a "blanket" is often interpreted (Emanuel, 2012: 95; updated in 2018).

The well-known "hockey stick" curve is depicted in Fig. 2.3 based on a thousand years of data for the northern hemisphere. The graph is not only widely cited for its conspicuous conformity to the shape of a hockey stick; it is also because of the careful integration of data from the past hundred years from instrumental records with the data from the previous 900 years from proxy data including tree rings and marine sediment. The dramatic increase during the twentieth century is obvious.

Because progress in scientific knowledge relies on replication and comparative results from different studies using different data-collection methods, it is useful to note the similar temperature patterns based on six different data sets covering 140 years, as presented in Fig. 2.4.

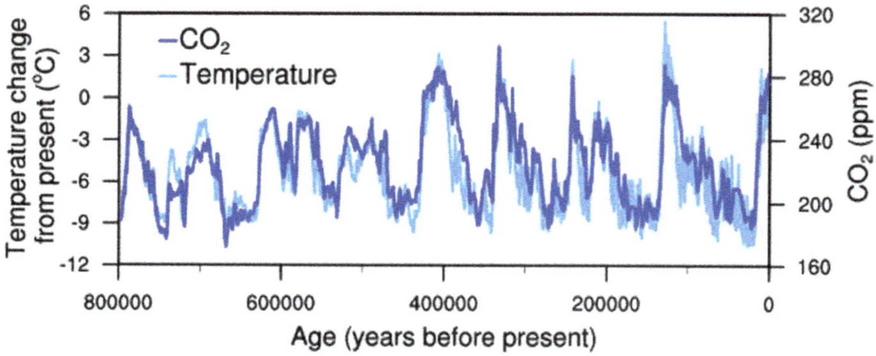

Fig. 2.2 Long-term relationship between temperature changes and CO_2 concentration level changes. *Source* US NOAA (2021), based on the EPICA Dome C ice core in Antarctica (Jouzel et al. 2007; Luthi et al. 2008)

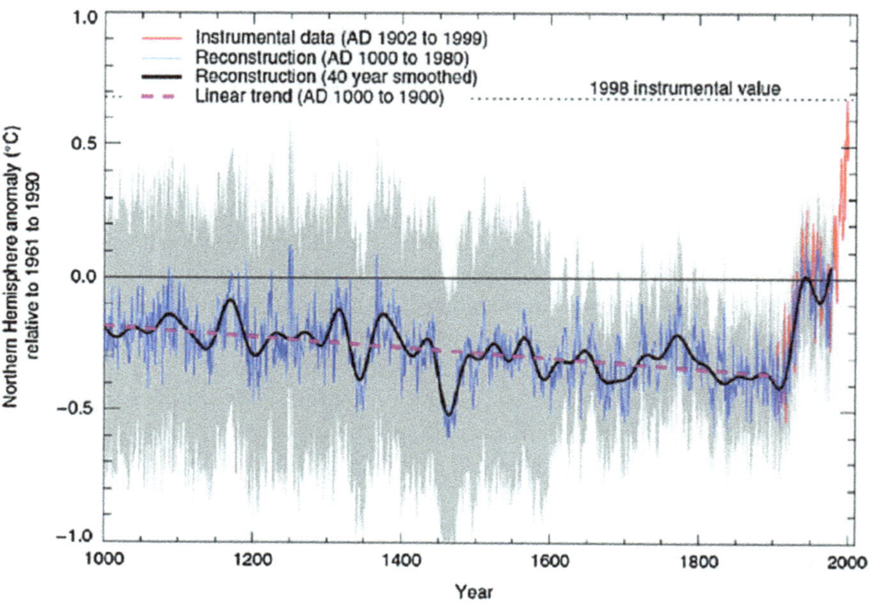

Fig. 2.3 The "hockey stick" curve of a thousand years of northern hemisphere temperatures, 1000–1999. *Source* Mann et al. (1999, 2001)

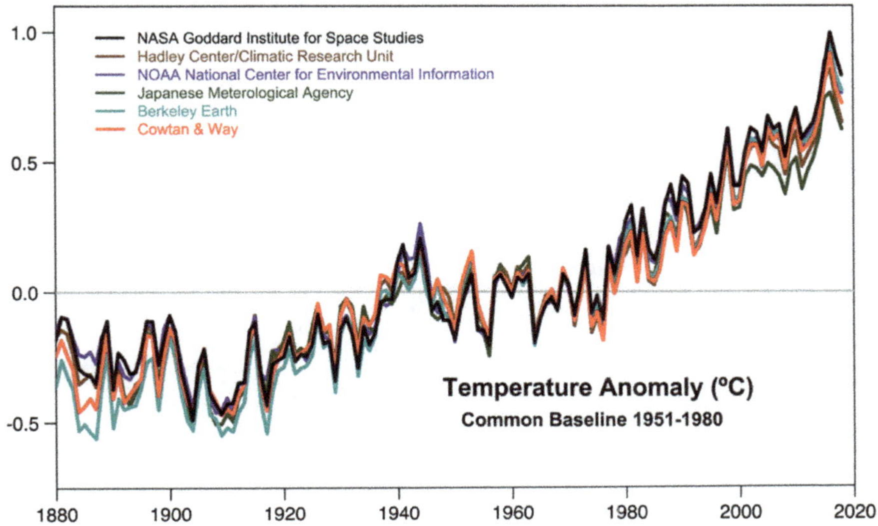

Fig. 2.4 Long-term global temperature changes. *Source* World Meteorological Organization (2020)

Another chart provides a succinct visual summary of recent global temperature changes and a series of projections until the end of the current century. Figure 2.5 is from the 2021–2022 reports of the Intergovernmental Panel on Climate Change (IPCC) prepared by hundreds of scientists from more than a hundred countries. Known as Assessment Reviews (ARs), the reports have been appearing every five to seven years since the first one in 1990. Each AR compiles the results of climate science studies since the previous AR. AR6 includes four volumes: a Synthesis Report; The Physical Science Basis; Impacts, Adaptation and Vulnerability; Mitigation of Climate Change.

The chart in Fig. 2.5 presents the empirically established global mean surface (GMS) temperature change during the period from 1950 to 2020, compared with the half century 1850–1900. By 2020, the GMS had increased by 1.1 °C (2.0 °F), compared with the average of the earlier period. The figure also includes projections into the future until 2100. As is commonly done in such projections—or "scenarios"—there are high, low and middle projections, with uncertainty bands—or "range estimates"—around the high and low projections. By 2050, the lower of the two low projections is above 1.5 °C, the higher of the two high projections is approaching 2.5 °C, and the middle projection is about 2 °C. These 2050 projections are thus in a dangerous range in terms of the frequency and severity of their effects.

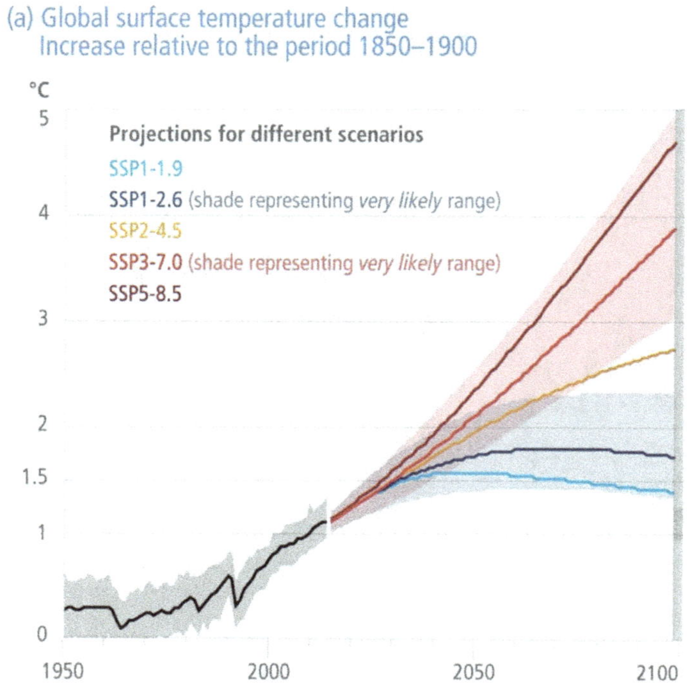

Fig. 2.5 Recent and future global surface temperatures. *Source* IPCC (2021a; used with permission)

2.4 Other Types of Emissions

2.4.1 Short-Lived and Long-Lived Emissions

In addition to carbon dioxide, there are many other types of chemicals in the emissions that cause climate change. They can be divided into *short-lived* and *long-lived* emissions depending on the molecules' average time in the atmosphere. As the accumulating scientific evidence about the rate and impacts of global warming has provoked a heightened sense of urgency among climate scientists, the distinction between short-lived and long-lived climate forcing agents has become more salient on the agendas of government policymakers, business executives, and other organizations with climate change interests.

In Table 2.1, the emissions are organized into five short-lived emissions, including black carbon as particulate matter, four long-lived emissions, and one mixed short-and-long category. The mixed category includes the large number of fluorocarbons, hydrofluorocarbons, and perfluorocarbons whose atmospheric lifetimes vary from a few days to thousands of years. The table presents introductory summaries of the atmospheric lifetimes and the global warming potentials (GWPs) of the diverse types of climate change forcing agents in addition to CO_2. GWP is the most widely used metric to indicate the potency *per tonne* of individual types of emissions, compared with carbon dioxide. An additional related metric is global temperature potential (GTP), which is indicative of a slightly different measure of the relative potency of a specific kind of emission. While GWP is a measure of the heat absorbed *over a given time period* relative to CO_2, GTP indicates the temperature change at the *end of a given period of time* relative to CO_2.

In Table 2.1 the large differences across types of emissions in their 20-year and 100-year GWPs are readily evident, and these numerical differences have significant implications for climate actions. One policy issue of increasing interest is that the relatively large 20-year GWPs of two major short-lived pollutants—methane and black carbon—are suggestive of the possibility of emission mitigation measures to achieve relatively quick greenhouse gas and particulate matter (ghg&pm) reductions.

There were new global emission concentration level records for all three of the principal greenhouse gasses in 2022 (NOAA, 2023). In 2022, the global concentration level of carbon dioxide (CO_2) was 419 ppm, which was 50% higher than pre-industrial level; the methane (CH_4) concentration level was 1912 ppb, which was more than 2½ times pre-industrial levels; and the nitrous oxide (N_2O) concentration level was 336 ppb, which was 24% above its pre-industrial level. The trends over decades for the three are indicated in Fig. 2.6.

In addition to methane and the other greenhouse gasses, black carbon particulate matter emissions have become more salient on the climate change action agenda as highly potent climate change forcing agents. Because they are Short-Lived Climate Pollutants (SLCPs), they offer opportunities for relatively quick reductions in emissions—on the order of days for black carbon and about a decade in the case of methane, compared with an average of a century for carbon dioxide. Furthermore, as the IPCC indicated in its AR6 report IPCC (2021), there is a potential of "win–win" mitigation policies that can reduce short-lived emissions such as methane and black carbon that are detrimental to both climate change and other features of air quality. These mitigation opportunities are summarized in Box 2.1.

Table 2.1 Types, atmospheric lifetimes and global warming potentials of greenhouse gas and particulate matter emissions

Emissions	Average atmospheric lifetime[a]	GWPs[a]: 20 years and 100 years	Other key features
Short-lived			
Aerosol clouds	Variable	Variable	Airplane contrails are a source
Black Carbon (BC)[b]	7–12 days[b]	3200[b] and 900	BC is particulate matter ($PM_{1.0}$), not a gas; it is among the three leading climate change forcing agents, with carbon dioxide and methane. Its impact on climate change occurs through *two* processes: one as an *aerosol* and the other as *deposition* on snow and ice. The latter is important because it reduces the albedo (reflective) effects of snow and ice. BC also causes respiratory, cardiovascular, asthma and other diseases, and damages agricultural crops. Diesel engines, forest clearing fires and wildfires are major sources
Methane (CH_4): fossil fuels	11.8 years[a]	83 and 30[a]	Flaring at oil wells and leaks from production, distribution and usage of methane are major sources. Methane is a precursor to ozone
Methane (CH_4): non-fossil fuels	11.8 years[a]	80 and 27[a]	
Organic Carbon (OC)		(Coolant)	Climate change *coolant*; common co-pollutant with BC in soot, but in much smaller volumes than BC, as emission from diesel engines
Ozone (O_3)[c]			Ozone is a *secondary* pollutant resulting from NO_x and VOC emissions

(continued)

Table 2.1 (continued)

Emissions	Average atmospheric lifetime[a]	GWPs[a]: 20 years and 100 years	Other key features
Long-lived			
Carbon Dioxide (CO_2)	100 years	[1.0] [1.0]	CO_2 is the leading climate change forcing agent; it is emitted by all transportation modes' internal combustion engines, and by fossil fuel electricity generation plants
Nitrous Oxide (N_2O)	109 years[a]	273[a] and 273	One of the four biggest overall contributors to global warming—long-lived and much more potent per tonne than CO_2 at 20 years and at 100 years. Nitrous oxide is a precursor to ozone
Sulphur Oxides (SO_x)		(Coolant)	Affects climate change as a *coolant*; damages human health
Sulfur Hexafluoride (SF_6)	3200 years	17,500 and 23,500	Extremely long-lived and potent
Mixture of short-lived and long-lived			
Fluorocarbons Hydrofluorocarbons Perfluorocarbons (CFCs, HFCs, HCFCs, PFCs)	Few days to thousands of years	Less than 1 to tens of thousands	Highly variable atmospheric lifetimes and GWPs, but all of them are among the most powerful climate forcing agents

Sources IPCC (2013, 2021a), Bond et al. (2013), European Commission (2021), US EPA (2021), US NASA (2021), Brewer (2017)

[a]A formal definition of GWP is: "the cumulative radiative forcing, both direct and indirect effects, over a specified time horizon resulting from the emission of a unit mass of gas related to some reference gas" (Bond et al. 2013). CO_2 is the base year for both 20-year and 100-year metrics. Reports of GWP data and lifetime data have evolved over time with additional research. Items marked "a" are from IPCC (2022a: Table 7.15)

[b]BC particulates are not only aerosols that float in the air; they also become depositions that fall onto snow and ice—and other surfaces—where they reduce the albedo (i.e. reflective) effect of snow and ice. The total climate change effect of BC thus includes an additional effect that gasses do not exhibit—namely the albedo effect of BC depositions (Bond et al. 2013)

[c]Voluntary Organic Compounds (VOCs) react with nitrogen oxides and carbon monoxide, as precursors to ozone. Carbon monoxide (CO) does not directly cause climate change, but it does contribute to climate change indirectly by increasing the amount of carbon dioxide and methane. Sources include all fossil fuels, and thus motor vehicles, planes, ships—and wildfires. VOCs are also a health hazard

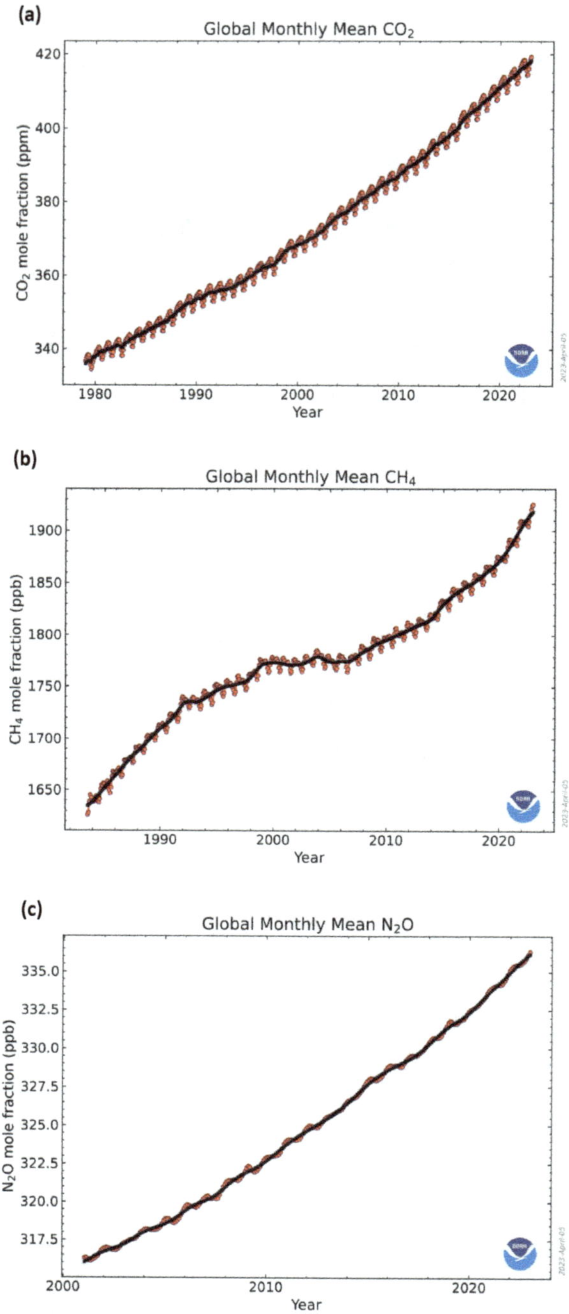

Fig. 2.6 Global mean monthly concentration levels for three principal greenhouse gasses. **a** Carbon dioxide. **b** Methane. **c** Nitrous oxide. *Source* NOAA Global Monitoring Laboratory in NOAA (2023)

Box 2.1: Links between limiting climate change and improving air quality
Most human activities, including energy production, agriculture, transportation, industrial processes, waste management and residential heating and cooling, result in emissions of gaseous and particulate pollutants that modify the composition of the atmosphere, leading to degradation of air quality as well as to climate change. These air pollutants are also short-lived climate forcers—substances that affect the climate but remain in the atmosphere for shorter periods (days to decades) than long-lived greenhouse gases like carbon dioxide. While this means that the issues of air pollution and climate change are intimately connected, air pollutants and greenhouse gases are often defined, investigated and regulated independently of one another in both the scientific and policy arenas. ... [S]ome short-term 'win–win' policies that simultaneously improve air quality and limit climate change include the implementation of energy efficiency measures, methane capture and recovery from solid-waste management and the oil and gas industry, zero-emissions vehicles, efficient and clean stoves for heating and cooking, filtering of soot (particulate matter) for diesel vehicles, cleaner brick-kiln technology, practices that reduce burning of agricultural waste, and the eradication of burning of kerosene for lighting.

Source: IPCC (2021b)

2.4.2 Scope 1, Scope 2, and Scope 3 Emissions

The standardized procedures for attributing emissions to specific sources include making a distinction among Scope 1, Scope 2, and Scope 3 emissions as indicated in Box 2.2.

Box 2.2: Definitions of Scope 1, 2 and 3 Emissions
Scope 1 emissions—direct emissions from sources owned or directly controlled by the organization. It uses the concepts of equity share or control (territorial, financial and operational) to establish Scope 1 emission responsibility.

Scope 2 emissions—indirect GHG emissions from purchased energy greenhouse gas emissions from the generation of purchased electricity, heat, cooling or steam consumed by the organization.

Scope 3 emissions—indirect GHG emissions that are a consequence of the organization's activities but arise from sources that are not owned or directly controlled by the organization. Scope 3 emissions include all attributable value chain GHG emissions not included in Scope 1 emissions or Scope 2 emissions.

Source: Excerpted and compiled by Thomas L. Brewer from International Standards Organization (2023, italics added; based on Protocol Corporate Accounting and Reporting Standard, in Greenhouse Gas Protocol, 2023).

2.4.3 Carbon Dioxide Equivalent (CO_2e) Emissions

Another widely used emission metric is carbon dioxide equivalent (CO_2e or CO_2-e), which is the GWP of a given kind of emission multiplied by the number of tonnes emitted. Thus, both GWP and CO_2e are indicators of the relative importance of types of emissions in their contributions to climate change. However, GWP is based only on the relative potency of a single tonne or other unit of volume. In contrast, CO_2e reflects the varying volumes of emissions as well as their GWP. CO_2e can thus be used for comparative aggregations that incorporate more than one kind of emission.

2.4.4 Earth's Carbon Budget

The concept of a *carbon budget* provides a single metric that encapsulates annual emissions relative to the earth's capacity to absorb the carbon in the emissions. There is a physical constraint to the amount of carbon the planet can absorb; when that constraint is reached, then global warming will continue until the goal of net zero emissions is achieved. Since oceans are a major sink for the emissions, the concept of the earth's carbon budget provides an encompassing notion of a trend that reflects the shortening of the time left before zero net emissions need to be reach. The concept and its implications have been stated succinctly:

> A carbon budget is a single number that encapsulates the finite limits of our planet's physical system and highlights the need to reach net zero.... [I]f we continue to release emissions on a net basis, the budget is breached and the temperature keeps rising (Carbon Tracker, 2022).

2.5 Effects of Emissions

2.5.1 Extreme Heat and Human Health

Extreme heat from global warming causes deaths and is otherwise a major threat to human health. As early as 1988, the city of Chicago experienced hundreds of deaths from a heat wave in the same summer as climate scientist Hansen (1988) warned the US Congress of the existence and consequences of global warming. More than a decade later, during the summer of 2003, extreme heat in Europe resulted in 30,000 deaths, including 15,000 in France. At that time, it was the hottest summer in the temperature records since 1851, and it was estimated at a confidence level greater than 90% that "human influence … at least doubled the risk of a heat wave of that magnitude (Stot et al. 2004). The summer of 2022 was also deadly in Europe, with more than 60,000 "heat-related deaths" from a summer heatwave in France, Germany, Spain and other European countries, (Ballester et al. 2023). Another study (Carbon Brief, 2021) of extreme heat events in many regions concluded that of the 122 events in the data base 92% were either made more likely or more severe by climate change. See Annex 1.1 for further information about the attribution of climate change as a cause of individual extreme weather events.

Although greenhouse gasses and particulates pose "global commons problems" that affect the entire planet, the effects are not distributed evenly among regions—a fact that is evident in the map in Fig. 2.7. The generally greater changes in over-land temperatures in the northern hemisphere, compared with the southern hemisphere, are evident, as are the most extreme world increases evident in the Arctic.

In the map, the dark areas of the Arctic are indications of its special place in the geography of the temperature effects of emissions. One consequence of Arctic heating is evidence (Chudley et al. 2019) that the amount of melted water under Greenland's glaciers that enters the ocean is much greater than previously realized and is thus leading to greater sea level rise than previously forecast (Zheng, 2022).

Climate scientists' interest in the Arctic extends to many other issues beyond Greenland's glacial melt and its effects on sea level rise, such as changes in the

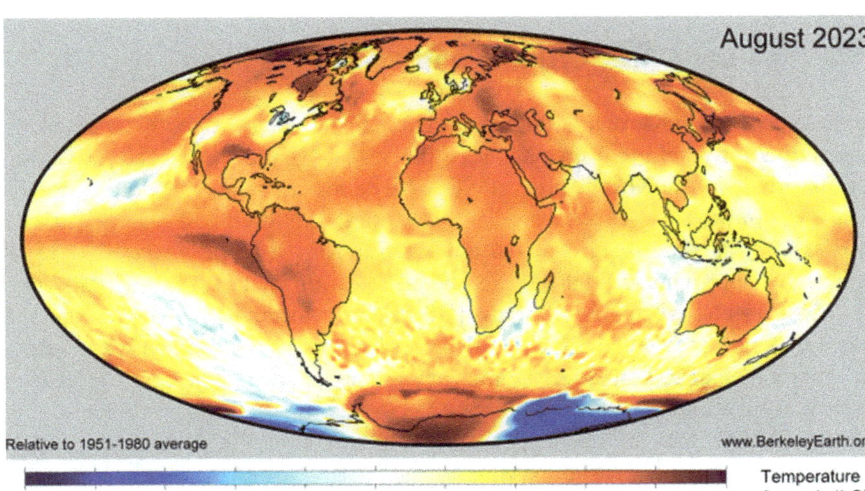

Fig. 2.7 Variations in temperature change among world regions. *Source* BerkeleyEarth.org (2022)

geographic patterns of the Atlantic ocean's gulf stream, and changes in the Arctic jet stream. Impacts of those changes occur on climate patterns as far away as the Sahara Desert in North Africa, and the cities and rural areas of the state of Texas and other states of the US. The jet stream (also called the "polar vortex") was a contributing factor to a damaging freeze in Texas in 2021 and a snow blizzard in Los Angeles in 2023 (Tigue, 2023).

Furthermore, an on-going Arctic-based topic of special interest is the discovery that temperature records for islands in the Barents Sea—which is within the Arctic circle—were the most rapid temperature increases in the world. Temperature increases between 5 and 7 times the global warming average were recorded on the Norwegian island Svalbard and the Russian island Franz Josef Land. The time period covered data from 40 years of data from 1981 to 2020 in a study by the Norwegian Meteorological Institute (2022).

2.5.2 Droughts, Crop Failures and Wildfires

The studies noted above (Carbon Brief, 2021) found that about two-thirds of the 69 droughts in the data base were more frequent or intense as a result of climate change. As experienced in many regions of the world in 2019–2022, the crop failures and wildfires associated with droughts often result in enormous economic and human losses over widespread areas. For instance, wildfires in Australia during its summer season from June 2019 until March 2020, after a prolonged drought, released 700 million tonnes of carbon dioxide, killed 33 people, destroyed 3000 homes, and burned more than 10 million hectares (the equivalent of 24.7 million acres) (Malapaty, 2021; Phillips & Nogrady, 2020; also see Calfire, 2022).

2.5.3 Sea Level Rise, Cyclones and Coastal Flooding

Coastal areas in numerous countries around the world are directly affected by coastal flooding resulting from cyclones ("hurricanes" in the US) and sea level rise. Because many of the world's biggest cities are in coastal areas, there are hundreds of millions of people who are directly exposed to these floods and storms. A study by Kirezci et al. (2020) summarized the exposure of the world's coastal areas as follows: "[A]pproximately 600 million people live in low elevation coastal zones [i.e. LECZs—coastal regions] less than 10 m above mean sea level (MSL). [The zones] generate approximately US$1 trillion of global wealth [annually]. … Both the environmental and socio-economic impacts associated with episodic coastal flooding can be massive."

Box 2.3 highlights the results of a study by the Union of Concerned Scientists (2018) of exposures to coastal flooding along the US east coast. Box 2.4 focusses on hurricane Ian in 2022.

Box 2.3: US east coast exposure to inundation

More than 300,000 of [US east coast homes in 2018], with a collective market value of about $117.5 billion …, are at risk of chronic inundation in 2045—a timeframe that falls within the lifespan of a 30-year mortgage issued today. Approximately 14,000 coastal commercial properties, currently assessed at a value of roughly $18.5 billion, are also at risk during that timeframe.

By the end of the century, homes and commercial properties currently worth more than $1 trillion could be at risk. This includes as many as 2.4 million homes—the rough equivalent of all the homes in Los Angeles and Houston combined—that are collectively valued today at approximately $912 billion.

The properties at risk by 2045 currently house 550,000 people and contribute nearly $1.5 billion toward today's property tax base. Those numbers jump to about 4.7 million people and $12 billion by 2100.

States with the most homes at risk by the end of the century are Florida, with about 1 million homes (more than 10% of the state's current residential properties); New Jersey, with 250,000 homes; and New York with 143,000 homes.

Source: Union of Concerned Scientists (2018)

Box 2.4: Climate change effects on hurricanes: the case of Ian

The conventional scale for hurricanes is an inadequate indicator of the damage a storm can do because the scale is based only on wind speed. However, particularly because of climate change, it is the flooding resulting from the storm surge and the rainfall that often do the most damage. US NOAA (2019) reported that "Storm surge flooding has accounted for nearly half of the deaths associated with landfalling tropical cyclones over the past fifty years." The storm surge tends to be higher because of the sea level rise that results from global warming since warmer water expands more. The rainfall also tends to be greater because warmer air holds more water.

All these features were evident in hurricane Ian that hit Florida in September 2022. The conventional hurricane rating based on wind speed put the hurricane in category 4 as it approached the coast and then only 1 after it landed, compared with a maximum possible 5; then, as its wind speed declined further, it became a *tropical storm* that passed across Florida and then the states of Georgia and South Carolina and North Carolina. In all, despite its relatively low ranking in conventional terms, it was the most damaging hurricane/tropical storm in a hundred years in Florida.

A report of the US NOAA (2022c) includes several record-breaking features of the storm as it moved across Florida and into South Carlina. Its storm surge was as high as 12–18 feet above ground level in some coastal areas. There were rainfalls of 17 inches in 12 hours and 17 inches in 24 hours. Early estimates of insured economic losses from the one storm were about US$ 47 billion. There were 87 human deaths. Of course, over time economic losses and human deaths are subject to revision, especially including estimates of uninsured losses.

However, it is important to note that climate change has not increased the overall frequency of hurricanes, but rather the severity of the most damaging ones. The damage, of course, also depends on the physical features of the terrain, built areas, and any storm control measures taken in the areas where the hurricanes hit. In the case of the west coast of Florida where hurricane Ian hit, there had been extensive building in low lying areas along the coast as the population increased in recent years. The largest city in the affected coast is Tampa, which has been one of the most vulnerable large cities in the world.

Finally, as for nomenclature, all these observations are the same for the storms that are called *cyclones* or *typhoons* in other parts of the world. They are all physically the same kind of storm but with different names.

Sources: BBC (2022), Miller et al. (2022), US NOAA (2019; 2022c)

2.6 Economic Costs and the Social Cost of Carbon

The economic impacts of climate change are increasingly of interest, as the annual costs continue to mount and as the costs are aggregated from local to national and global levels. Computations of the economic costs of climate change are also used increasingly at the level of specific policy issues; at that level of analysis, cost–benefit computations are undertaken to compare the economic costs of implementing regulatory policies that reduce emissions compared with the benefits to society of the avoided-costs by reducing climate change. Thus, a key element of such analyses is commonly referred to as the "social cost of carbon" (SCC).

Basic SCC concepts were developed by economists in the 1980s and 1990s, beginning with Nordhaus (1982). Over time, the key concepts have been refined, as have the data and models used to compute the costs and benefits. At the same time, even as the core analyses and their results depend on economic reasoning and information technologies for processing the data, there have also been politically driven changes in key numbers—including in the US, in particular. Furthermore, there have been lawsuits in the US about the appropriate institutionalized procedures for using the SCC computational results in the budgeting process (Brewer, 2022).

2.7 Sources of Emissions

2.7.1 Countries

The three biggest national emitters of carbon dioxide—China, the US and India—together have contributed about half the global total, albeit with year-to-year variations. China's share (about 33%) of the global total has been more than twice the US share (13%), and the US share has been about twice India's (7%). The next three—Russia, Japan, Germany—have together contributed about another 10%, respectively 5%, 3% and 2%.

These and other data about big emitters are displayed in Table 2.2. Of course, the enormous differences in the six countries' populations mean that their rankings in terms of emissions per capita are quite different; the US is thus higher than any of the other countries. There are yet other differences in the rankings according to emissions per a standardized amount of GDP, such as per thousand US dollars. For this metric, China is the most carbon intensive economy, while Germany is the least carbon intensive.

2.7.2 Economic Sectors

The distinctive issues posed by economic sectors and their industries have been studied extensively, and they have become increasingly contentious. A few summary observations are nevertheless appropriate here. An underlying phenomenon in several of the sectors is the importance of fossil fuels (coal, oil and natural gas) as the primary sources of the emissions of individual sectors. This has been particularly true of the electric power sector, the transportation sector, the building sector, and some industries in the industry sector. These sectors and industries have long been among the principal sectoral sources of greenhouse gas and particulate emissions.

Table 2.2 Carbon dioxide emissions of major emitting countries

Country	Amount (2020)			
	% of World	Billion tonnes	Tonnes per capita	Tonnes per thousand US$ of GDP
China	32.5	11.7	8.2	0.51
US	12.6	4.5	13.7	0.23
India	6.7	2.4	1.7	0.29
Russia	4.7	1.7	11.6	0.43
Japan	3.0	1.1	8.4	0.21
Germany	1.8	0.6	7.7	0.15

Source European Commission EDGAR (2021), also see European Union Copernicus (2022)

2.8 Conclusion

Climate science has made much progress in tracking the many types of emissions that have been causing global warming over a long period of time. Significant patterns and trends in the emissions are well documented. Although projections into the future are inevitably accompanied by many uncertainties, a range of quantitative scenarios can establish low, middle and high estimates of the rates of emissions decades into the future.

It is the a*ccumulated net concentration levels* of carbon dioxide and other climate change forcing agents that are the key to understanding the amount of warming and its consequences. It is also important to recognize that in addition to carbon dioxide there are other kinds of long-range and short-range emissions that are significant contributors to warming patterns and trends. For instance, nitrous oxides are important long-term agents with average atmospheric lifetimes of about a century or more, while methane gas and black carbon particulate matter are highly potent short-term agents with average atmospheric lifetimes, respectively, of only about a decade and ten days.

As a consequence of the increasing concentration levels of these and other climate change forcing agents in recent decades, there have been significant increases in the frequency and severity of extreme weather events. In the early decades of the twenty-first century, the destructive effects have become more evident, as a result of severe hurricanes as well as floods, droughts, crop failures, and wildfires. Although sea-level rise is less conspicuous, its effects are already catastrophic along coasts where there are major population concentrations; sea level rise thus poses extraordinary long-term risks to human lives and economies.

Annex 2.1: Attribution of the Causes of Extreme Weather Events

Attribution analysis is based on a comparison of the results of climate model runs based on (1) a scenario of a period of time without global warming and (2) a scenario of the same period of time with global warming. The period is often as long as about 150 years to cover the industrial revolution era to the present (Cho, 2021).

The World Weather Attribution (WWA) collaboration (2021; also see Sjoukje et al. 2020) has developed a standardized methodology for selecting, analyzing and reporting probabilistic studies of causal connections between climate change and individual extreme weather events. The sequence of steps is summarized as follows:

1. the trigger: which studies to perform,
2. the event definition: which aspect of the extreme event were most relevant,
3. observational trend analysis: how rare was it and how has that changed,
4. climate model evaluation: which models can represent the extreme,
5. climate model analysis: what part of the change is due to climate change,
6. hazard synthesis: combine the observational and model information,
7. analysis of trends in vulnerability and exposure, and
8. communication of the results.

Source: World Weather Attribution (2021)

Among the early studies of the attribution of an extreme weather event to climate change was a paper about the "Human Contribution to the European Heat Wave of 2003" (Stott et al. 2004). Since then, there have been numerous studies such as Frame et al. (2020), Gori et al. (2022), Keellings and Hernández (2019).

The summary results of more than 350 attribution studies of 405 extreme weather events are reported in Table 2.3.

Extreme events in the US and Europe in 2021, 2022 and 2023 increased awareness of "compound extreme" events, such as a combination of extreme heat, extreme droughts and extreme floods all happening in the same place at the same time. Although the co-occurrence of heat and droughts is intuitive, the co-occurrence with flooding may be counter intuitive. However, climate scientists have known of the phenomenon and warned about it for many years, and they can explain it. With increasing surface temperatures, more water vapor evaporates into the atmosphere. As atmospheric temperature increases, it increases the atmosphere's capacity to hold moisture; it is a basic fact of atmospheric physics. As a result, more powerful storms with more water cause more flooding.

Table 2.3 Results of 350 attribution studies of 405 extreme weather events

Type of event (number)	Percent made more likely or more severe by human-caused climate change
Droughts (69)	65%
Rainfall or flooding (81)	58%
Extreme heat (122)	92%
Other (133)	na
Total extreme events (405)	70%

Source Carbon Brief (2021)

Annex 2.2: Warming in the Arctic

The Arctic is sometimes referred to as the climate change's symbolical equivalent of the *canary in the mine*—an analogy with the use of the caged canaries' deaths in coal mines to indicate dangerous levels of carbon monoxide for the human workers in them. Arctic temperatures have increased by about four times the global mean increase in recent decades (Rantanen et al. 2022).

This annex discusses the types and sources of emissions that are causing the warming, and the global consequences as well as the local and regional consequences. There is an emphasis on black carbon emissions because of their relatively large and distinctive contributions to the warming trend. There is also an emphasis on the causal mechanisms and consequences of shifts in the "jet stream" and sea level rise.

In the Arctic, black carbon and methane, as well as carbon dioxide, are all major climate change agents. Climate scientists have long known that black carbon is an important climate change forcing agent; the first Assessment Review in 1990 by the International Panel on Climate Change (IPCC) noted its existence (IPCC, 1990; also see Andreae et al. 1988). However, because black carbon is not a gas, it has often been left out of policymaking discourse about the sources of global warming (Brewer, 2016, 2019). One of the frontiers of climate change analysis for many years has been the integration of black carbon data and related equations in mathematical models that describe patterns and trends in climate change—and project them into the future (University Corporation for Atmospheric Research, 2022).

The photograph in the annex Fig. 2.8 reveals the extent of black carbon (BC) deposits on the snow and ice surfaces in the Arctic—particularly in Greenland. The BC on Greenland is a major cause of melting glaciers, which discharge enormous volumes of fresh water into the nearby Arctic Sea; that fresh water, in turn, changes Atlantic Ocean salinity, temperature and currents. Greenland's glacial melt also contributes significantly to global sea-level rise (Voosen, 2020). All of these effects are felt in the US, especially along the Atlantic coast from Maine to Florida.

Black carbon deposits on Arctic Sea ice have different consequences from those on Greenland's glaciers because ice floating in the sea displaces its water-equivalent volume. Nevertheless, the extent of sea ice has decreased, and the possibilities for maritime shipping traffic to traverse the Arctic have thereby increased (Brewer, 2016, 2019).

Black carbon is not the only climate forcing agent in the Arctic region. Carbon dioxide and methane are also present. As in other regions of the world, carbon dioxide is present as a globally mixed, long-lived greenhouse gas, and it thus has continuing warming effects year after year.

Figure 2.9 in the Annex shows the substantial decrease of the extent of Arctic Sea ice in recent years during the period November to March, with its peak in the northern hemisphere in the spring and its annual low point in the fall. The unusually low extent in 2020–2021 is evident in the solid blue line at the bottom. Concomitant increases in the volumes of cruise ship and cargo ship traffic are

Fig. 2.8 Black carbon deposits on Greenland's glaciers. Credit: ICCT

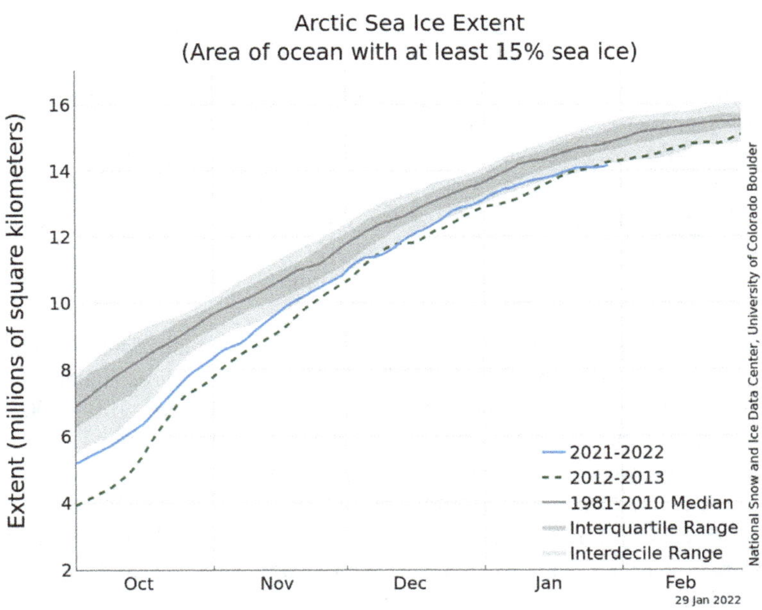

Fig. 2.9 Extent of arctic sea ice during November-March. *Source* US National Snow and Ice Data Center (2022)

already evident. Increased Arctic ship traffic adds to the black carbon burden in the region—which is the result of a combination of local sources such as wood stoves and quite distant sources such as coal-fired electric power plants in China (Yamineva, 2020).

There are global consequences of the warming of the Arctic. The consequences include sea level rise, changes in the Atlantic Gulf stream, climate changes resulting from changes in the Arctic "jet stream" and the potential of a global "tipping point" from arctic methane leaks (Arctic Council, 2022; Intergovernmental Panel on Climate Change, 2019).

Annex 2.3: Annual State of the Global Climate Reports

The World Meteorological Organization (WTO) issues an annual *State of the Global Climate* report that contains extensive data and analysis of recent developments. Its report for 2022 is available in Arabic, Chinese, English, French, Russian, and Spanish (WMO, 2023).

Its "key messages for the year" include:

The global mean temperature in 2022 was 1.15 °C above the 1850–1900 average. The years 2015–2022 were the eight warmest in the 173-year instrumental record.

Concentrations of the three main greenhouse gases—carbon dioxide, methane and nitrous oxide—reached record highs in 2021, the latest year for which consolidated global values are available (1984–2021). The annual increase in methane concentration from 2020 to 2021 was the highest on record. Real-time data from specific locations show that levels of the three greenhouse gases continued to increase in 2022.

Around 90% of the energy trapped in the climate system by greenhouse gases goes into the ocean.

Global mean sea level continued to rise in 2022, reaching a new record high for the satellite altimeter record (1993–2022). The rate of global mean sea level rise has doubled between the first decade of the satellite record (1993–2002, 2.27 mm per year) and the last (2013–2022, 4.62 mm per year).

In East Africa, rainfall has been below average in five consecutive wet seasons, the longest such sequence in 40 years.

Record-breaking rain in July and August led to extensive flooding in Pakistan. There were at least 1700 deaths, and 33 million people were affected, while almost 8 million people were displaced. Total damage and economic losses were assessed at US$ 30 billion.

Record-breaking heatwaves affected China and Europe during the summer. In some areas, extreme heat was coupled with exceptionally dry conditions. Excess deaths associated with the heat in Europe exceeded 15,000 in total across Spain, Germany, the United Kingdom, France and Portugal.

Source: Excerpted and compiled by Thomas L. Brewer from WMO (2023).

This and related content of the report led the WMO Secretary-General to conclude that: "This report shows that, once again, greenhouse gas concentrations in the atmosphere continue to reach record levels – contributing to warming of the land and ocean, melting of ice sheets and glaciers, rising sea levels, and warming and acidifying of oceans" (Taalas, 2023; in WMO, 2023).

Also see regional reports from the WHO:

State of the Climate in Africa
State of the Climate in Asia
State of the Climate in Europe
State of the Climate in Latin America and the Caribbean
State of the Climate in the South-West Pacific.

Questions to Ponder

1. Assume that you are a reporter for a major news source and that your editor has asked you to write a story that highlights the key issues about the future of climate change. After reading this chapter, what would your headline and themes be?
2. If you want to determine whether carbon dioxide in the atmosphere is an increasing problem, why does it matter whether you use changes over time in the annual rates of carbon dioxide emissions or changes over time in the concentration levels?
3. Why is carbon dioxide generally considered the most important greenhouse gas?
4. What are examples of short-term climate pollutants? How—and how much—do they differ from carbon dioxide?
5. Do you agree with the quote at the beginning of the chapter? Why or why not?

References

Andreae, M., et al. (1988). Bio-mass burning emissions and associated haze layers above Amazonia. *Journal of Geophysical Research*, 93(2), 1509–1527. https://dash.harvard.edu/bitstream/handle/1/14117818/Biomass%20burning%20emissions%20and%20associated%20haze%20layers%20over%20Amazonia.pdf?sequence=1. Accessed July 1, 2022.

Arctic Council. (2022). Black carbon and methane. Accessed at https://www.arctic-council.org/about/task-expert/egbcm/ on November 18, 2022.

Ballester, J., et al. (2023). Heat-related mortality in Europe during the summer of 2022. *Nature Medicine*, *29*, 1857–1866. https://www.nature.com/articles/s41591-023-02419-z#:~:text=On%20average%20over%20the%2035,distribution%20of%20summer%20mean%20temperatures. Accessed October 28, 2023.

BBC. (2022). Hurricane Ian—Maps and images showing destruction. https://www.bbc.com/news/world-us-canada-63078606. Accessed May 29, 2023.

BerkeleyEarth.org. (2022). Whats New: October 2022 Update. https://berkeleyearth.org/whats-new/?cat=temperature-updates. Accessed December 16, 2022.

Bond, T. C., et al. (2013). Bounding the role of black carbon in the climate system: A scientific assessment. *Journal of Geophysical Research: Atmospheres, 118*, 5380–5552. https://doi.org/10.1002/jgrd.50171. Accessed February 20, 2022.

Brewer, T. (2016). Proposal for an Arctic Black Carbon (ABC) Agreement. Presentation at workshop on Marine Black Carbon Emissions: BC Control Strategies, sponsored by the International Council on Clean Transport (ICCT) in collaboration with Environment and Climate Change Canada, Vancouver.

Brewer, T. (2017). Black carbon problems in transportation: Technological solutions and governmental policy solutions. Working Paper, MIT Center for Energy and Environmental Policy Research, July 2017. Available at www.ceepr.mit.edu.

Brewer, T. (2019). The Arctic Ocean's melting ice: Institutions and policies to manage black carbon. In P. G. Harris (Ed.), *Climate change and ocean governance* (Chap. 15). Cambridge University Press.

Brewer, T. (2022). *Transforming US Climate Change Policies: 2021 and Beyond*. Policy Brief. Springer. https://link.springer.com/book/9783030997175.

Briley, L., et al. (2021). A Practitioner's Guide to Climate Model Scenarios. Great Lakes Integrated Sciences and Assessments (GLISA). Ann Arbor, MI. glisa-info@umich.edu. Accessed December 12, 2022.

Brundtland, G. (1988). World Meteorological Organization (WMO) and United Nations Environment Programme (UNEP). The Changing Atmosphere: Implications for Global Security Implications. WMO/0MM-No. 710. Toronto, Canada, 27–30 June 1988. https://wedocs.unep.org/20.500.11822/29980. Accessed July 9, 2021.

CalFire, Stats and Events. https://www.fire.ca.gov/stats-events/. Accessed March 8, 2022.

Carbon Brief. (2021). Mapped: How climate change affects extreme weather around the world. https://www.carbonbrief.org/mapped-how-climate-change-affects-extreme-weather-around-the-world. Accessed February 20, 2022.

Carbon Tracker. (2022). https://carbontracker.org/carbon-budgets-where-are-we-now/. Accessed March 6, 2022.

Carr, H., & Great Lakes Integrated Sciences and Assessments (GLISA). (2021). A Practitioners' Guide to Climate Model Scenarios. https://glisa.umich.edu/wp-content/uploads/2021/03/A_Practitioners_Guide_to_Climate_Model_Scenarios.pdf. Accessed August 10, 2022.

Carr, H., Rheingans, P., & Schumann, H. (2014). Similarity explorer: A visual inter-comparison tool for multifaceted climate data. In *Eurographics Conference on Visualization* (Vol. 33(3)). https://www.srs.fs.usda.gov/pubs/ja/2014/ja_2014_poco_001.pdf. Accessed August 3, 2022.

Cho, R. (2021). Attribution science: Linking climate change to extreme weather. *Phys Org*, October 5, 2021. https://phys.org/news/2021-10-attribution-science-linking-climate-extreme.html. Accessed February 20, 2022.

Chudley, T. R., et al. (2019). Supraglacial lake drainage at a fast-flowing Greenlandic outlet glacier. *Proceedings of the National Academy of Sciences, 116*(51), 25468–25477. https://doi.org/10.1073/pnas.1913685116. Accessed February 22, 2022.

Emanuel, K. (2012). *What we know about climate change.* Cambridge, MA, 2918; also see updated edition in 2018.

Emanuel, K. (2023). Climate primer: Climate science, risk and solutions. https://climateprimer.mit.edu/. Accessed May 15, 2023.

European Commission. (2021). EDGAR—Emissions Database for Global Atmospheric Research. https://edgar.jrc.ec.europa.eu/. Accessed June 7, 2021.

European Space Agency. (2021). https://www.esa.int/Applications/Observing_the_Earth/Copernicus/Sentinel-5P/Mapping_methane_emissions_on_a_global_scale. Accessed April 11, 2021.

European Union, Copernicus. (2022). Climate Change Service. https://climate.copernicus.eu/about-us. Accessed March 23, 2022.

Foitzheim, N., Majka, J., & Zastrozhnov, D. (2021). Methane release from carbonate rock formations in the Siberian permafrost area during and after the 2020 heat wave. *Proceedings of the National Academy of Sciences (PNAS), 118*(32). https://doi.org/10.1073/pnas.2107632118. Accessed January 29, 2022.

Frame, D., Wehner, M., & Noy, I., et al. (2020). The economic costs of Hurricane Harvey attributable to climate change. *Climatic Change, 160*, 271–281. https://doi.org/10.1007/s10584-020-02692-8. Accessed February 21, 2022.

Francis, J., & Skific, N. (2015). Linking rapid Arctic warming to mid-latitude weather. Patterns. *Philosophical Transactions of the Royal Society, 13 July 2015.* Accessed at https://royalsocietypublishing.org/doi/full/10.1098/rsta.2014.0170 on November 18, 2022.

Frankhauser, S., et al. (2022). The meaning of net zero and how to get it right. *Nature Climate Change, 12*, 15–21. https://doi.org/10.1038/s41558-021-01245-w. Accessed May 1, 2022.

Friedlingstein, P., et al. (2022). Global Carbon Budget 2021. https://essd.copernicus.org/preprints/essd-2021-386/essd-2021-386.pdf. Accessed March 6, 2022.

Gori, A., Lin, N., Xi, D., & Emanuel, K. (2022). Tropical cyclone climatology change greatly exacerbates US extreme rainfall–surge hazard, 03 February 2022. https://www.nature.com/articles/s41558-021-01272-7#Abs1. Accessed February 4, 2022.

Greenhouse Gas Protocol. (2023). Standards. https://ghgprotocol.org/about-us. Accessed April 22, 2023.

Hansen, J. E. (1988). The greenhouse effect: Impacts on current global temperature and regional heat waves. Testimony to U.S. Senate, Committee on Energy and Natural Resources, June 23, 1988. Washington, DC. https://www.sealevel.info/1988_Hansen_Senate_Testimony.html. Accessed July 8, 2021.

Hook, L. (2022). How is the jet stream connected to simultaneous heatwaves across the globe? *Financial Times,* 21 July. https://www.ft.com/content/bb9ecca3-c914-489d-af7d-06df6010d1ea. Accessed July 27, 2022.

Intergovernmental Panel on Climate Change (IPCC). (1990). *Climate change: The IPCC scientific assessment.* Cambridge University Press.

Intergovernmental Panel on Climate Change (IPCC). (2000). *Special Report on Emissions Scenarios (SRES).* Cambridge University Press. https://www.ipcc.ch/site/assets/uploads/2018/03/emissions_scenarios-1.pdf Accessed August 10, 2022.

Intergovernmental Panel on Climate Change (IPCC). (2001). Third Assessment Review (TAR), Working Group I, Climate Change 2001: The Scientific Basis. Fig. 2.20, p. 134. https://www.ipcc.ch/report/ar3/wg1. Accessed March 23, 2021.

Intergovernmental Panel on Climate Change (IPCC). (2013). Fifth Assessment Review, Working Group I, Climate Change 2013: The Scientific Basis. https://www.ipcc.ch/report/ar5/wg1. Accessed March 20, 2020.

Intergovernmental Panel on Climate Change (IPCC). (2019). Special Report on the Ocean and Cryosphere in a Changing Climate: Summary for Policymakers. Accessed at www.ipcc.ch/srocc/ on December 16, 2022.

Intergovernmental Panel on Climate Change (IPCC). (2021a). Summary for Policymakers, *Climate Change 2021: The Physical Science Basis.* Contribution of Working Group I to the Sixth Assessment Report of the Intergovernmental Panel on Climate Change. https://www.ipcc.ch/report/ar6/wg1/downloads/report/IPCC_AR6_WGI_SPM_final.pdf, 16 February 2022. Accessed September 7, 2022.

Intergovernmental Panel on Climate Change (IPCC). (2021b). *Climate Change 2021: The Physical Science Basis.* Contribution of Working Group I to the Sixth Assessment Report of the Intergovernmental Panel on Climate Change. https://www.ipcc.ch/report/sixth-assessment-report-working-group-i/. Accessed June 15, 2022.

Intergovernmental Panel on Climate Change (IPCC). (2022a). *Climate change 2022: Impacts, adaptation and vulnerability.* The Working Group II contribution to the Sixth Assessment Report. https://www.ipcc.ch/report/sixth-assessment-report-working-group-ii/. Accessed June 15, 2022.

Intergovernmental Panel on Climate Change (IPCC). (2022b). *Climate Change 2022: Mitigation of Climate Change.* Annex III: Scenarios and Modelling Methods. The Working Group III contribution to the Sixth Assessment Report. https://www.ipcc.ch/report/sixth-assessment-report-working-group-3/. Accessed August 11, 2022.

International Energy Agency (IEA). (2023). World Energy Outlook. https://www.iea.org/topics/world-energy-outlook. Accessed on October 9, 2023.

International Institute for Sustainable Development (IISD). (2022). A warming arctic is a warning signal for the world. Accessed at https://www.iisd.org/system/files/2022-04/still-one-earth-arctic.pdf on December 16, 2022.

International Standards Organization (ISO). (2023). Net Zero Guidelines. https://www.iso.org/obp/ui/en/#iso:std:iso:iwa:42:ed-1:v1:en. Accessed April 22, 2023.

Jouzel et al. (2007). Orbital and millennial antarctic climate variability over the past 800,000 years. *Science*, 317(5839), 793–796. https://doi.org/10.1126/science.1141038. Accessed February 23, 2022.

Keellings, D., & Hernández Ayala, J. J. (2019). Extreme rainfall associated with Hurricane Maria over Puerto Rico and its connections to climate variability and change. *Geophysical Research Letters, 46*, 2964–2973. https://doi.org/10.1029/2019GL082077. Accessed February 21, 2022.

Keller, R. (2019). Europe's killer heat waves are a new norm. The death rates shouldn't be. *Washington Post*, 26 July 2019. https://www.washingtonpost.com/outlook/2019/07/26/europes-killer-heat-waves-are-new-norm-death-rates-shouldnt-be/. Accessed July 9, 2021.

Kirezci, E., et al. (2020). Projections of global-scale extreme sea levels and resulting episodic coastal flooding over the 21st century. *Nature, Scientific Reports*, 10. https://www.nature.com/articles/s41598-020-67736-6. Accessed March 8, 2022.

Liu, Z., et al. (2020). Near-real-time monitoring of global CO_2 emissions reveals the effects of the COVID-19 pandemic. *Nature Communications, 11*, 5172. https://doi.org/10.1038/s41467-020-18922-7

Lüthi et al. (2008). Temperature change and carbon dioxide change. https://www.ncdc.noaa.gov/global-warming/temperature-change. Accessed August 6, 2021.

Mallapaty, S. (2021). Australian bush fires belched out immense quantity of carbon. *Nature*, 15 September 2021. https://www.nature.com/articles/d41586-021-02509-3. Accessed March 8, 2020.

Mann, M., et al. (1999). Northern hemisphere temperatures during the past millennium. *Geophysical Research Letters, 26*, 759–762.

Mann, M., et al. (2001). Proxy-based reconstructions of hemispheric and global surface temperature variations over the past two millennia. *Proceedings of the National Academy of Sciences (PNAS), 105*(36), 13252–13257.

MIT. (2023). Climate Portal. https://climate.mit.edu/. Accessed May 15, 2023.

Miller et al. (2022). Hurricanes are becoming more dangerous. Here's why. https://www.cnn.com/interactive/2020/12/us/hurricanes-climate-change/index.html. Accessed May 29, 2023.

NOAA. (2023). Greenhouse gasses continued to increase rapidly in 2022. https://www.noaa.gov/news-release/greenhouse-gases-continued-to-increase-rapidly-in-2022. Accessed April 16, 2023.

Nordhaus, W. (1982). How fast should we graze the global commons? *American Economic Review, 72*(2), 242–46. https://econpapers.repec.org/article/aeaaecrev/v_3a72_3ay_3a1982_3ai_3a2_3ap_3a242-46.htm. Accessed March 8, 2022.

Østergaard, P., Andersen, A., & Sorknæs, P. (2022). The business-economic energy system modelling tool energyPRO. *Energy, 257*, 124792. https://www.sciencedirect.com/science/article/pii/S0360544222016954. Accessed August 3, 2022.

Phillips, N., & Nogrady, B. (2020). The race to decipher how climate change influenced Australia's record fires, *Nature*, 23 January. https://www.nature.com/articles/d41586-020-00173-7. Accessed March 8, 2020.

Rantanen, M., et al. (2022). The Arctic has warmed nearly four times faster than the globe since 1979. *Communications Earth and Environment*. https://www.nature.com/articles/s43247-022-00498-3 Accessed August 11, 2022.

Schreuder, C., & Stein, S. (1995). Heat's toll worse than believed. *Chicago Tribune*, 21 September. https://www.chicagotribune.com/news/ct-xpm-1995-09-21-9509210301-story.htm. Accessed July 9, 2021.

Science Daily. (2021). Arctic methane release due to melting ice is likely to happen again, March 22, 2021. https://www.sciencedaily.com/releases/2021/03/210322135221.htm. Accessed January 29, 2022.

Sjoukje, P., et al. (2020). A protocol for probabilistic extreme event attribution analyses. *Advances in Statistical Climatology, Meteorology and Oceanography*, *6*, 177–203. https://doi.org/10.5194/ascmo-6-177-2020. Accessed February 20, 2022.

Stott, P., et al. (2004). Human contribution to the European heatwave of 2003. *Nature, 432*, 610–614. https://doi.org/10.1038/nature03089. Accessed February 21, 2022.

Taalas, P. (2023). Forward in World Meterological Organization (WMO) (2023) State of the Global Climate 2022. https://library.wmo.int/doc_num.php?explnum_id=11593. Accessed April 22, 2023.

Tigue, K. (2023). How climate change and the polar vortex influenced this week's harsh winter storms, today's climate. https://mail.google.com/mail/u/0/?ik=f940d45579&view=pt&search=all&permthid=thread-f%3A1758744136417698010&simpl=msg-f%3A1758744136417698010. Accessed on February 24, 2023.

Union of Concerned Scientists. (2018). Underwater: Rising Seas, Chronic Floods, and the Implications for US Coastal Real Estate, 18 June 2018. https://www.ucsusa.org/resources/underwater. Accessed March 8, 2022.

UK Met Office. (2021). Atmospheric carbon dioxide at record high levels despite reduced emissions in 2020. https://www.metoffice.gov.uk/research/news/2021/record-co2-levels-despite-lower-emissions-in-2020. Accessed July 30, 2021.

United Nations Environment Programme (UNEP). https://www.unep.org/publications-data. Accessed June 15, 2022.

US Environmental Protection Agency (EPA). (2021). Climate Change Indicators: U.S. Greenhouse Gas Emissions. Fig. 3. U.S. Greenhouse Gas Emissions per Capita and per Dollar of GDP, 1990–2019. https://www.epa.gov/climate-indicators/climate-change-indicators-us-greenhouse-gas-emissions. Accessed March 7, 2022.

US National Aeronautics and Space Administration (NASA). (2021). Global Climate Change. https://climate.nasa.gov/. Accessed June 7, 2021.

US National Oceanic and Atmospheric Administration (NOAA). (2019). Climate Change. https://www.climate.gov. Accessed July 10, 2021.

US National Oceanic and Atmospheric Administration (NOAA). (2020). Climate change: Atmospheric carbon dioxide. By Rebecca Lindsey. https://www.climate.gov/news-features/understanding-climate/climate-change-atmospheric-carbon-dioxide. Accessed July 10, 2021.

US National Oceanic and Atmospheric Agency (NOAA). (2021a). NOAA Research News. Carbon dioxide peaks near 420 parts per million at Mauna Loa observatory. https://research.noaa.gov/article/ArtMID/587/ArticleID/2764/Coronavirus-response-barely-slows-rising-carbon-dioxide. Accessed July 30, 2021.

US National Oceanic and Atmospheric Agency (NOAA). (2021b). NOAA Research News. Carbon dioxide peaks near 420 parts per million at Mauna Loa observatory. https://research.noaa.gov/article/ArtMID/587/ArticleID/2764/Coronavirus-response-barely-slows-rising-carbon-dioxide. Accessed July 30, 2021.

US National Oceanic and Atmospheric Administration (NOAA). (2022a). Carbon dioxide now more than 50% higher than pre-industrial levels. https://www.noaa.gov/news-release/carbon-dioxide-now-more-than-50-higher-than-pre-industrial-levels. Accessed August 7, 2022.

US National Oceanic and Atmospheric Administration (NOAA). (2022b). NOAA Climate.gov. https://www.climate.gov/. Accessed June 10, 2022.

US NOAA (2019).

US NOAA. (2022). Hurrican Ian's Path of Destruction. https://www.nesdis.noaa.gov/news/hurricane-ians-path-of-destruction. Accessed May 29, 2023.

US NOAA. (2022c). Hurrican Ian's Path of Destruction. https://www.nesdis.noaa.gov/news/hurricane-ians-path-of-destruction. Accessed May 29, 2023.

US National Snow and Ice Data Center. (2022). Arctic Sea Ice Extent, 29 January 2022. http://nsidc.org/arcticseaicenews/. Accessed January 29, 2022.

University Corporation for Atmospheric Research (UCAR). (2022). Greenland's ice is melting. https://scied.ucar.edu/learning-zone/climate-change-impacts/greenlands-ice-melting. Accessed January 29, 2022.

Voosen, P. (2020). Seas are rising faster than ever, *Science*, November 18. https://www.sciencemag.org/news/2020/11/seas-are-rising-faster-ever. Accessed on January 26, 2021.

Weart, S. (2003). The discovery of rapid climate change. *Physics Today*, *56*(8), 30. https://doi.org/10.1063/1.1611350. Accessed July 8, 2021.

Woods Hole Oceanographic Institute. (2023). Carbon Dioxide in Seawater. https://www.whoi.edu/know-your-ocean/ocean-topics/how-the-ocean-works/ocean-chemistry/ocean-acidification/carbon-dioxide-in-seawater/. Accessed April 23, 2023.

Woollings, T. (2020). *Jet stream: A journey through our changing climate*. Oxford University Press.

World Meteorological Organization (WMO). (2020). Global Temperature Figures. https://library.wmo.int/doc. Accessed April 22, 2023.

World Meterological Organization (WMO). (2023). State of the Global Climate 2022. https://library.wmo.int/doc_num.php?explnum_id=11593. Accessed April 22, 2023.

World Weather Attribution. (2021). Pathways and Pitfalls in extreme event attribution. https://www.worldweatherattribution.org/pathways-and-pitfalls-in-extreme-event-attribution/. Accessed February 19, 2022.

Yamineva, Y. (2020). Reducing China's black carbon emissions: An arctic dimension. The Arctic Institute. https://www.thearcticinstitute.org/reducing-china-black-carbon-emissions-arctic-dimension/. Accessed January 29, 2022.

Zheng, W. (2022). Glacier geometry and flow speed determine how Arctic marine-terminating glaciers respond to lubricated beds. *The Cryosphere, 16*, 1431–1445. https://doi.org/10.5194/tc-16-1431-2022. Accessed April 28, 2022.

Resources for Keeping Up with Developments in Climate Science and Climate Conditions

Carbon Tracker. https://carbontracker.org/

Emanuel, K. *Climate primer: Climate science, risk and solutions*. https://climateprimer.mit.edu/

European Commission (EC). EDGAR—Emissions Database for Global Atmospheric Research. https://edgar.jrc.ec.europa.eu/

European Union (EU), Copernicus. Climate Change Service. https://climate.copernicus.eu/

Intergovernmental Panel on Climate Change (IPCC) (Periodic reports on many issues). https://www.ipcc.ch/

International Energy Agency (IEA) (Annual and monthly reports on energy efficiency and other climate-related topics). https://www.iea.org/

MIT. Climate Portal. https://climate.mit.edu/

UK Met Office (Climate change developments). https://www.metoffice.gov.uk/about-us/press-office/news/weather-and-climate/

United Nations Environment Programme (UNEP), World Environment Situation Room. https://wesr-climate.unepgrid.ch/

US National Aeronautics and Space Administration (NASA). Global Climate Change. https://climate.nasa.gov/

US National Oceanic and Atmospheric Administration (NOAA). https://www.climate.gov/

UK MET Office

World Bank. https://climateknowledgeportal.worldbank.org/

World Meteorological Organization (WMO). State of the Global Climate [annual]. https://public.wmo.int/en/our-mandate/climate/wmo-statement-state-of-global-climate

World Weather Attribution [project]. https://www.worldweatherattribution.org/

The Big Emitters and The Most Vulnerable People

3

> What we are now doing to the world, by degrading the land surfaces, by polluting the waters and by adding greenhouse gases to the air at an unprecedented rate … . It is mankind and his activities which are changing the environment of our planet in damaging and dangerous ways.
>
> UK Prime Minister Thatcher (1989)
>
> United Nations General Assembly Speech
>
> The most vulnerable people bear the brunt of climate change impacts yet contribute the least to the crisis. As the impacts of climate change mount, millions of vulnerable people face greater challenges in terms of extreme events, health effects, food security, livelihood security, water security, and cultural identity.
>
> World Bank, Social Dimensions of Climate Change (2022)

3.1 Introduction

This chapter focuses on two groups of countries: those that are the biggest emitters in terms of overall volume of the emissions and their climate impacts, and those that are the most vulnerable to the effects. Virtually all people are directly or indirectly vulnerable to the effects of climate change, and hundreds of millions are already being directly affected by it. There is, of course, variability in exposure to the effects of climate change among countries and groups of people within countries; further, the locational variability interacts with the kind of climate change effect.

T. Brewer, *Climate Change*, https://doi.org/10.1007/978-3-031-42906-4_3

The two groups of countries have been included in the same chapter for several reasons. Two of the biggest emitters—China and India—are also among the most vulnerable to its effects. They are among the most vulnerable countries partly because of their large populations, but also because of their extensive low-lying coastal areas, large densely populated urban areas around seaports, and desert and semi-desert areas. There is an emphasis generally on low income countries, especially but not only low-lying islands; however, there are millions of people in large urban areas in low-lying coastal regions in many large countries that are vulnerable to sea level rise and cyclones. Millions more in rural regions are vulnerable to long-term droughts, recurrent episodes of extreme heat, and destructive floods. These vulnerabilities are evident in high-income as well as low-income countries. Also, international financial assistance programs involve both groups as donor countries or recipient countries, and the ethical issues also involve both kinds of countries as well as groups within them.

Section 3.2 presents a comparative overview of the emissions of the big emitters. In Sect. 3.3, detailed profiles of the big emitters' policies are discussed. Given the inherent importance of the global impacts of the big emitters' emissions, the chapter presents comparative analyses of the countries' emission *mitigation* policies. The EU27 is included in the analysis as an entity representing 27 countries because it has its own collective policies; as an entity, it is often ranked among the world's top four emitters.

Section 3.4 shifts the focus to especially vulnerable countries. It begins with a list of major extreme weather events of various types that affected developing countries in Asia and Africa during the two year period 2021–2022. Special attention is given to Small Island Developing States (SIDS) and other islands that are particularly vulnerable in Sect. 3.5. Multiple on-going crises in East Africa that have been triggered by climate change also receive special attention in Sect. 3.6.

International financial "loss and damage" compensation programs are the subject of Sect. 3.7, and ethical issues are the subject of Sect. 3.8. Section 3.9 summarizes the chapter.

3.2 Comparative Overview of Emissions

In recent years, the three biggest emitters, as measured by carbon dioxide emissions—China, the US and India—have together contributed half the global total. China's share of the global total has been more than twice the US share, and the US share about twice India's. The next three—Russia, Japan, Germany—have together contributed only another 11%.

In Table 3.1, section a includes the EU and five individual countries, which collectively contribute two-thirds of the world total. In section b, the remainder of the list includes 11 countries, each of which contributes more than one percent of the world total. Germany is included in the EU27 in section a and separately in section b because of its high ranking as a single country.

Table 3.1 Big emitting countries with CO_2 emissions more than 1% of the world total (2020)

Countries	CO_2 Gigatonnes	CO_2 Gigatonnes—percent World Total	CO_2 Tonnes—Per Capita	*Ratio of* CO_2 Tonnes Per Capita *to* world average (4.62)	CO_2 Tonnes—Per US$ 000 GDP[b]
Section a					
China	11.7	32.5	8.2	1.8	0.507
US	4.5	12.6	13.7	3.0	0.228
EU27[a]	2.6	7.3	5.9	1.3	0.141
India	2.4	6.7	1.7	0.4	0.285
Russia	1.7	4.7	11.6	2.5	0.432
Japan	1.1	3.0	8.4[c]	1.8[c]	0.213
Section b					
Iran	0.7	1.9	8.3[c]	1.8[c]	0.661
Germany[a]	0.6[c]	1.8[c]	7.7	1.7	0.150
South Korea	0.6[c]	1.7[c]	12.1	2.6	0.284
Saudi Arabia	0.6[c]	1.6[c]	17.0[d]	3.7	0.382
Indonesia	0.6[c]	1.6[c]	2.1	0.5	0.181
Canada	0.5[c]	1.5[c]	14.4	3.1	0.311
Brazil	0.5[c]	1.3[c]	2.1	0.5	0.151
South Africa	0.4[c]	1.2[c]	7.4	1.6	0.640
Mexico	0.4[c]	1.1[c]	3.0	0.6	0.177
Australia	0.4[c]	1.1[c]	15.2[d]	3.3	0.309
Turkey	0.4[c]	1.1[c]	5.2	1.1	0.169

Sources Compiled by the author from data in European Commission, Joint Research Centre (2021)

[a] Germany is listed separately because of its high ranking as an individual country in section b, but it is also included in the EU27 in section a

[b] Purchasing Power Parity (PPP)

[c] Minor inconsistencies are due to rounding from the original source

[d] Others in the top ten in per capita emissions but not in the table because their individual country emissions are relatively small are four oil producing Persian Gulf countries: Bahrain, Kuwait, Qatar, UAE

Note that these are CO_2 emissions, which have an average atmospheric life of about a century. CO_2 emissions are the most consequential over the longer term and the most common focus of attention among indicators of emission macro patterns and trends. The patterns and consequences of major short-lived methane gas and black carbon particulate matter emissions—which are not in the table—are different because of their much shorter average atmospheric lives but much greater warming potency per tonne

Rankings in terms of emission volumes per capita are quite different. Four countries—Australia, Canada, Saudi Arabia, and the US—have per capita emissions three or more times greater than the world average. Rankings in terms of emissions volumes relative to national economic production are also different. Four countries—Canada, Iran, Russia, and South Africa—all have significant fossil fuel industries.

An important clarification of these rankings is that they are based on production, not consumption—sometimes referred to, respectively, as measures of "territorial" and "carbon footprint" emissions. When exported goods and imported goods are included, the profiles of some individual countries change significantly. A well know example is the UK, which has been a net importer of carbon intensive goods since it closed most of its coal industry; its consumption-based emissions have been as much as 45% more than its production-based emissions. Meanwhile, in the US, consumption-based emissions have been about 10% more than production-based emissions. In contrast, China's consumption-based emissions have been as low as 12% less than its production-based emissions in recent years, and India's consumption-based emissions have been about 10% less than its production-based emissions.

These production-consumption differences are important because a reliance on production-based data is contrary to a long-standing principle of environmental economics—namely the "consumer pays" principle. There are on-going efforts to adjust international comparisons among countries' emissions records by adjusting for their exports and imports, but the data requirements for making accurate adjustments are formidable. Nevertheless, approximate adjustments are indicative of the kinds of differences in countries' emissions patterns that emerge among major economies: although there are exceptions, wealthy countries tend to be net importers and thus consumers of emissions-intensive goods that are produced and exported by less wealthy countries. China is, of course, a net exporter of goods to many countries, including the US, UK, Germany, and many other European countries as well as high income countries in other regions. These trade-related issues are pertinent to international legal agreements in the context of countries' emission inventories that are reported according to the production-based requirements of UN Framework Convention on Climate Change (UNFCCC) guidelines.

Another issue involving international trade concerns trade in services by maritime shipping and by aviation, which are treated differently in terms of both applicable international legal agreements and emissions reporting systems. These two industries have been formally agreed in international negotiations to be separated from the UNFCCC national emission reporting processes and to be addressed instead by two specialized UN agencies, the International Civil Aviation Organization (ICAO) and the International Maritime Organization (IMO). The emissions in those two industries, therefore, are not included in individual countries' emission inventories for the UNFCCC, but instead in the reporting systems of the two specialized agencies.

3.3 Profiles of The Big Emitters

3.3.1 European Union (EU-27)

The EU has strived to be a world leader on climate action issues for many years, and it has succeeded in important ways. Emissions have been trending downward for several decades. Yet, the EU remains a big emitter as a single entity that includes 27 member countries.

Emission Patterns and Trends

A report by Eurostat (2022) summarizes the history of the EU trends of CO_2 equivalent emissions during the three-decade period from 1990 to 2019.

> In 2019, EU GHG emissions were over 1 billion tonnes of CO_2 equivalent *lower* [italics added] than in 1990. This corresponds to a 24 percent reduction compared with 1990 levels, which is more than the [previously agreed] EU reduction target of 20 percent by 2020. … GHG emissions were below 1990 levels in 22 Member States. The largest reductions, over 50 percent, were recorded in Estonia, Romania, Lithuania and Latvia.

Figure 3.1 traces actual CO_2 equivalent emissions from the onset of the Emissions Trading System (ETS) in 2005 until 2020, and then projects the levels until 2030.

Note that black carbon (BC) particulate matter emissions are not included in the CO_2-e data in Fig. 3.1. The exclusion of BC was a common practice in climate science analyses for many years because BC emissions occur as particulate matter,

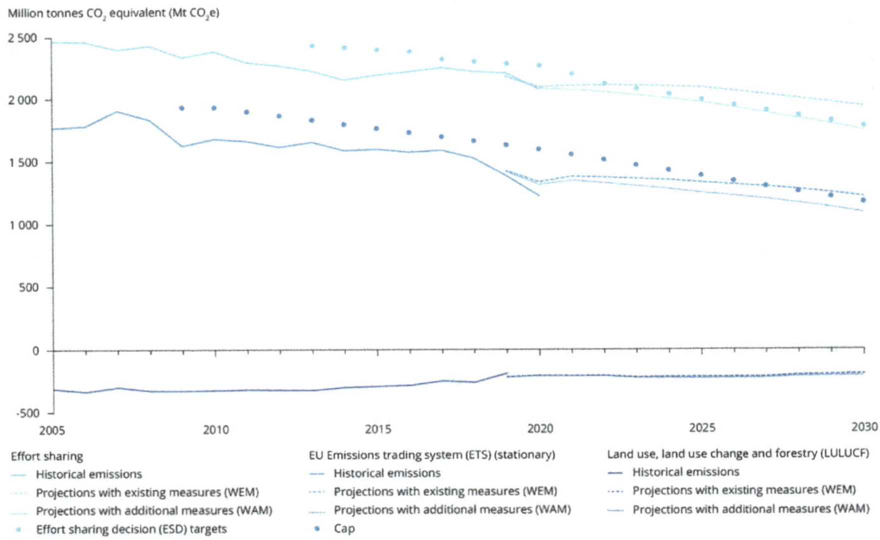

Fig. 3.1 EU-27 CO_2-e emissions: actual for 2005–2020; projections for the subsequent period. *Source* EEA (2022)

not a gas. However, BC emissions are among the three principal global warming forcing agents along with methane and carbon dioxide, and they are increasingly recognized as potential targets for short-term mitigation actions.

However, at the sector and industry-specific levels of analysis, the inclusion of BC emissions is more common, and in some industries its inclusion is essential to an understanding of the nature of the emissions and their impacts. Because of the mitigation co-benefits for human health and food production resulting from reduced BC emissions, it is also important to include BC in analyses of mitigation measures. This is especially important in the transportation sector, where diesel engines in motor vehicles and maritime shipping are major sources of BC emissions (Comer & Osipova, 2021).

Effects and Adaptation Policies

European countries have suffered from most of the common effects of climate change, including floods, heat waves, declines in agricultural production and sandstorms in recent years. Over the longer-term, sea level rise poses potentially catastrophic effects in coastal areas—where there have already been large-scale adaptation policies at the national and local levels in some countries. Dust storms that originate in northern Africa and travel across the Mediterranean into Spain and other countries have become more frequent in recent years, but there is not much that countries in Europe can do except warn residents to stay inside.

Among the EU institutions, there has been active development of mitigation policies for more than three decades—both in its international relationships and in its internal policymaking.

Mitigation Policies: Chronological Overview

The EU has aspired for years to be a world leader in addressing climate change issues, and it has succeeded in many respects. During the period 1990–2012 emissions declined by 19%, while over the same period, GDP *increased* 44%. Emissions and GDP were thus "decoupled". During the period 1990–2020 (before the UK's exit), the EU's CO_2-e emissions declined by 31% (EEA, 2022; Eurostat, 2022).

However, although EU emissions have generally declined in most of the major economic sectors, the transportation sector has been an important exception, and it remains a topic of much contention. In the electric power sector, the continuing use of coal in several countries also remains a contentious issue, as does the future of nuclear power.

As for policy initiatives, the EU was the world leader in the establishment of an Emissions Trading System (ETS) in 2005. New policies developed in 2019–2023 are expanding the industries covered and otherwise making the ETS more effective with new regulations. Box 3.1 presents a list of landmark events in the evolution of the EU's climate change policymaking over three decades from the early 1990s to the early 2120s. Box 3.2 highlights provisions in the wide-ranging changes in the "Fit for 55" program.

Box 3.1: Landmark Events in EU Policymaking on Climate Change Issues
1992: EU was an early supporter and then signatory to the UNFCCC.

1996: EU Environment Council, consisting of ministers of EU members, endorsed limiting average global surface temperature increase to less than 2 °C (3.6 °F) since the pre-industrial period.

1997: EU ratified the Kyoto Protocol of the UNFCCC, thereby agreeing to an 8% CO_2 reduction by 2008–2012 compared with 1990.

2005: EU established the European Emission Trading System (ETS), which has been a significant cause for the decline in emissions. During the period 2008–2015 the ETS reduced CO_2 emissions by about 1.2 billion tonnes, or almost half of the EU's Kyoto Protocol target (Bayer & Aklin, 2020).

2007: EU Council, consisting of the heads of state and government of EU members, endorsed limiting the mean global surface temperature increase to less than 2 °C (3.6 °F) compared with the level in the pre-industrial period.

2009: A "climate and energy package" of six legislative proposals was approved by the EU Parliament and Council. It provided for expanding and otherwise revising the ETS; establishing an Effort Sharing Decision for the EU members to meet new 2020 emission goals; promoting renewable energy sources; specifying and reducing the use of fossil fuels; establishing emission performance standards for cars and light-duty vehicles; and geological storing of carbon dioxide.

2015: EU agreed to the Paris Agreement at the UNFCCC COP21 meeting.

2019–2020: European Commission presented "The European Green Deal," consisting of more than a dozen wide-ranging proposals for diverse actions in many sectors, with a goal of net-zero ghg emissions in 2050—and with a new interim goal of a reduction of 50–55% by 2030 compared with 1990. During 2019 the European Parliament and the European Council both endorsed the 2050 goal.

2021: The Commission proposed a "Fit for 55" array of policies, many of which were specific proposals for implementation of the European Green Deal, but there were also more ambitious proposals as well. The "55" in the title of the package of proposals refers to the new 55% reduction target for ghgs by 2030 compared with 1990. This statement of the 2030 target was precise and definite about the 55%, compared to the previously considered 50–55% range. The substance of the package to achieve the 55% target and to begin progress toward the 2050 target was elaborated in a series of legislative proposals developed by the Commission for consideration by the Parliament and Council.

2022: The Parliament and the Council approved the "Fit for 55" legislative package after prolonged negotiations over key provisions, especially concerning transportation and energy. The resulting legislation was wide-ranging and detailed (see Box 3.2).

Sources: (European Commission, 2019a, 2019b, 2021a, 2021b, 2021c; European Union, Official Journal, 2021).

Box 3.2: Summary of the EU's "Fit for 55" Provisions

- *Emissions Trading System (ETS)*—Lower the overall emissions cap, phase out free emission allowances for aviation within the EU, include shipping emissions.
- *Carbon Border Adjustment Mechanism (CBA) for the ETS*—Some products will be covered by CBAs "to ensure that ambitious climate action in Europe does not lead to carbon leakage."
- *New ETS for fuel distribution for road transport and buildings*—Revenues should be allocated to reduce the social impact on vulnerable households, micro-enterprises and transport users.
- *Effort Sharing Regulation*—Reductions targets for many sectors based on individual members' GDPs per capita, with adjustments for coal efficiency.
- *Regulation on Land Use, Forestry and Agriculture*—Overall EU reductions of CO_2 emissions to 310 million tonnes by 2030. Plant 3 billion trees by 2030.
- *Renewable Energy Directive*—Produce 40% of energy from renewable sources by 2030.
- *Energy Efficiency Directive*—Nearly double the annual energy saving obligations of members. The public sector must renovate 3% of its buildings per year until 2030.
- *CO2 emissions standards for cars and vans*—Average emissions of new cars will be reduced 55% by 2030 and 100% by 2035 compared with 2021. All new cars registered as of 2035 will be zero-emission.
- *Alternative Fuels Infrastructure Regulation*—EU member states must expand electric charging capacities consistent with zero-emission car sales. Major highways must have electric charging points every 60 km and hydrogen refueling points every 150 km.
- *ReFuelEU Aviation Initiative*—Jet fuel at EU airports have to blend increasing levels of sustainable aviation fuels.
- *FuelEU Maritime Initiative*—There will be a limit on the ghg content of energy used by ships entering EU ports.
- *Energy Taxation Directive*—Remove current low tax rates that encourage fossil fuel use; promote clean technologies with reduced tax rates.
- *Social Climate Fund*—Create new fund for financing citizens' investments in energy efficiency, new heating and cooling systems, and cleaner mobility. Over the period 2025–2032 the fund will provide 144.4 billion

euros derived from a combination of the new building and road fuels ETS revenues, the EU budget and matching funds from member states.

Sources: European Commission (2021a, 2021b, 2021c), European Parliament (2022)

After more than three decades of initiating and refining a broad range of climate change policies, by the mid-2020s the EU has become a world leader in many respects. Although other countries in other regions have developed their own ETS systems, the EU's ETS has been the largest in terms of the volume of emissions being reduced and the volume of credits being traded. See Annex 2.1 for details of the evolution of the Emission Trading System (ETS) policies over four phases during 2005–2029.

3.3.2 Germany

Germany is not only important as one of the world's big emitters, it is also important for its influence as the country with the biggest national population and economy in the EU. This section thus briefly complements the previous section, with its emphasis on the EU as a single entity. Here, the focus is on differences among EU member countries on climate change issues and policy responses. Germany has a record as sometimes a leader and sometimes a laggard within the EU.

Emissions

Germany's total emissions declined from 1251 million tonnes in 1990 to 746 million tonnes CO_2-e in 2022—a 40% decrease. The biggest decline was in the energy sector, which decreased from more than 400 million tonnes to about 250.

As for the future, it has set targets of a 65% reduction by 2030 compared with 1990, and 88% by 2040 compared with 1990. By 2050, it intends to be carbon neutral. These and other elements of Germany's national climate laws are consistent with EU laws. As with EU laws, Germany's emissions are inclusive of all ghg emissions, unlike many countries in other regions of the world, where emissions target often refer only to CO_2.

Among the four big national emitters, Germany has the best overall ratings by Climate Action Tracker (2023) in the comparative table reported in Sect. 3.3.6.

Policy Conflicts in the EU

However, Germany has been at odds with some of the other EU members and the EU Commission on EU policies concerning the transportation and energy sectors ("Charlemagne" of the Economist 2023). The transportation conflict concerned the EU proposal to end the sale of new internal combustion cars by 2035, which the German government opposed in order to protect its large automobile industry.

A compromise was reached whereby internal combustion cars could still be sold but only if they were climate neutral. In the energy sector, the conflict was over Germany's energy subsidies of 200 billion Euros (USD 211 billion equivalent) for corporations and households to offset the effects on energy prices in Germany because of the war in the Ukraine. Although it was not an EU program, many EU members objected to it because they could not match it and there was thus politically damaging domestic resentment in many countries.

Climate Change Threats in Germany

As a northern European country, the effects of climate change and perceptions of them in Germany tend to be different from those in southern Europe, especially along the northern coast of the Mediterranean Sea. Heat is a greater threat to human health in the south. There are also some similarities, of course. All the major seaports of Europe, wherever they are located, are exposed to the effects of sea level rise. At the same time, there are variations among and within countries in exposure to sea level rise, depending on the elevation above sea level of coastal areas. Floods are a threat everywhere there are large rivers, which includes much of Europe, even in mountainous areas. There were major floods in 2021 (when 189 people died) and more floods again in 2022, and 2023 (when some cities were more than half destroyed).

3.3.3 US

In the US, the testimony of James Hansen before a Senate Committee in the summer of 1988 marked the beginning of a new era for climate change issues in the US because of his prominence as a widely-respected climate scientist in the US National Aeronautics and Space Administration (NASA). In his testimony, he said that he wanted ...

> ... to draw three main conclusions. Number one, the earth is warmer in 1988 than at any time in the history of instrumental measurements. Number two, the global warming is now large enough that we can ascribe with a high degree of confidence a cause and effect relationship to the greenhouse effect. And number three, our computer climate simulations indicate that the greenhouse effect is already large enough to begin to affect the probability of extreme events such as summer heat waves (Hansen, 1988).

That same summer a heat wave resulted in more than a hundred deaths in Chicago. Of course, US emissions and concentration levels had been increasing before 1988. The basic facts, patterns and trends in US emissions rates, concentration levels, and temperatures are well established. Figure 3.2 traces the carbon dioxide concentration levels for more than half a century in Mauna Loa, Hawaii. Because carbon dioxide is a well-mixed, globally-distributed gas, these concentration levels are of course the result of carbon dioxide emissions from other counties in addition to the US. Nevertheless, it is a widely recognized source of CO_2 concentration data, and it is well-known for the regularity of the annual seasonal pattern around the

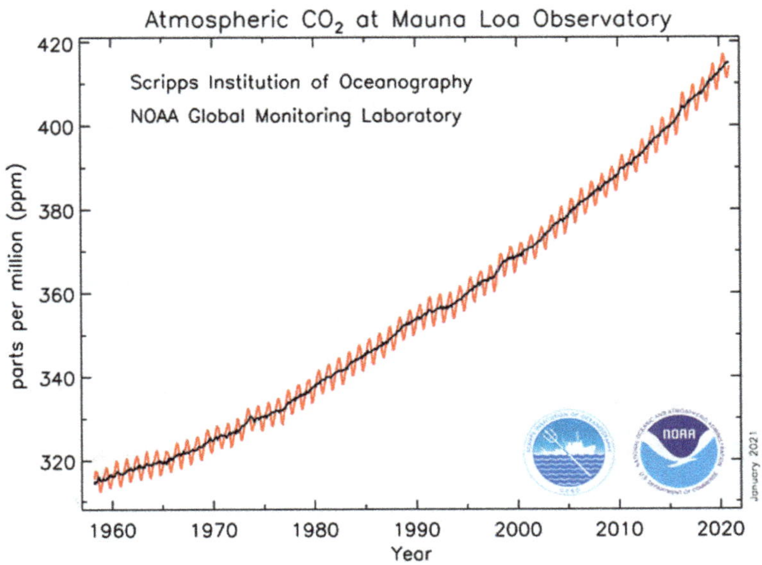

Fig. 3.2 Atmospheric concentration levels of carbon dioxide. *Source* US NOAA (2022)

steady long-term increase. The seasonal pattern reflects the decrease in the northern hemisphere's summer months when forestry and other foliage absorb carbon dioxide, as compared with their reduced abundance in the winter months.

In Fig. 3.3 the total and sectoral sources of US CO_2-e emissions over 30 years are presented. Electricity generation and transportation have together produced roughly half of the emissions. These proportions are likely to change in coming years as solar and wind installations continue to grow and as electric vehicles become increasingly common.

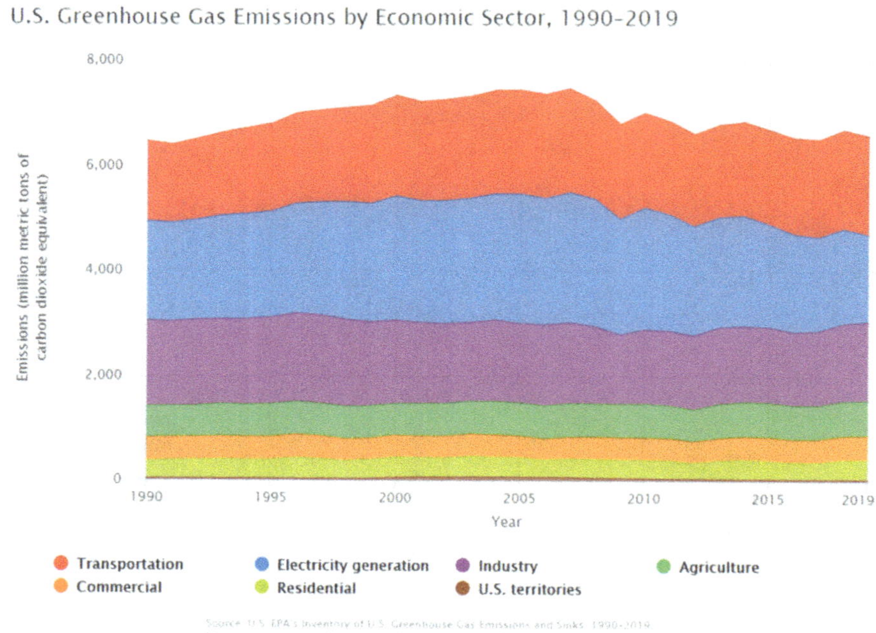

Fig. 3.3 Sectoral sources of US emissions. *Source* US EPA (2022)

3.4 Mitigation Policy

The history of US policymaking on climate change mitigation issues at the national level has been marked by a combination of significant changes in presidential approaches as administrations have changed. In addition, partisan Congressional differences have often prevented action (Brewer, 2022). However, the passage of climate change legislation in 2022 was a landmark event in the history of US climate change policymaking. After many months of internal party negotiations with a key Senator from a coal state, the legislation that was originally named the "Build Back Better" bill when introduced by President Biden, had its name changed to the "Inflation Reduction Act" (see Box 3.3).

Box 3.3: Major US Climate Change Legislation Passed in 2022
The total budget amount was US$ 369 billion. The new legislation was also estimated potentially to achieve a 40% reduction in US emissions by 2030, short of the 50% of the officially declared target. The new legislation, though, established major new spending over a ten-year period on programs in many sectors—including energy, transportation, industry, buildings, agriculture, and forestry.

The energy programs include subsidies for producers and consumers. The biggest programs in monetary terms are a US$ 30 billion program that accelerates "manufacturing of solar panels, wind turbines, batteries, and critical minerals processing," and another US$ 30 billion program "for states and electric utilities to accelerate the transition to clean electricity." An energy program of US$ 27 billion is "to support deployment of technologies to reduce emissions, especially in disadvantaged communities." A regulatory provision affecting the energy sector is a new program of fines to reduce methane leaks at oil and gas facilities.

As for the transportation sector, there are programs for: seaports (US$ 3 billion), heavy-duty vehicles (US$ 1 billion), US Postal Service purchase of zero-emission vehicles (US$ 3 billion); and "grants to retool existing auto manufacturing facilities to manufacture clean vehicles" (US$ 2 billion).

The agriculture sector receives over US$ 20 billion "to support climate smart agricultural practices," and forestry programs, including conservation and urban tree planting that are budgeted at US$ 5 billion.

Pricing Carbon in the US

Many state governments had already begun to take significant action on climate change issues before the presidential election of 2016, and they increased their efforts after that presidential election year as the Trump administration canceled and reversed many Obama administration climate initiatives. In fact, a regional emission trading system had already been established by a group of northeastern states. Known as the Regional Greenhouse Gas Initiative (RGGI), it is focused on reducing carbon dioxide emissions in the power sector. It is the first mandatory market-based program in the United States to reduce greenhouse gas emissions. Its members are the eleven states of Connecticut, Delaware, Maine, Maryland, Massachusetts, New Hampshire, New Jersey, New York, Rhode Island, Vermont, and Virginia. Meanwhile on the west coast, California had already created a cap-and-trade program in 2013. At the outset, the program included six greenhouse gases from industrial and electricity-producing facilities. In 2015, transportation fuels and natural gas were added. While California's population and economy have grown, greenhouse emissions have declined. The target established by the state legislature is that the 2030 emission level will be 40% below the 1990 level.

In any case, the future of climate change policies at the state and national government levels as well as local levels will of course depend in part public opinion.

Age and Partisan Differences in Public Opinions

Several surveys of US public opinion have found consistent patterns in the differences among age groups. As for the size and the political significance of age differences, it is age differences within the Republican Party that are especially interesting for their potential impact on the future of US policies. The results of a Pew survey taken in May 2021 revealed both the partisan differences and age differences in US adults' attitudes. The numeric specifics of the differences are

Table 3.2 Age and partisan differences in climate change attitudes in the US

Question						
Age in 2021	18–24	25–40	41–56	57–75	Difference: Youngest-Oldest	All adults
Year born[a]	1997–2012	1981–1996	1965–1980	1956–1964		
All adults (%)	77	75	67	64	13	70
Democrats (%)	92	91	91	93	1	92
Republicans (%)	55	52	40	35	20	42
Difference: Dem.-Rep. (%)	+ 37	+ 39	+ 51	+ 58	–	+ 50

[a] Breakdowns for those born in 1928–1945 and aged 76–93 in 2021 are not included in the original source table.
Source Compiled by the author from Pew (2021)

displayed in Table 3.2. The survey results reaffirm often-observed partisan differences: There was a 50% difference between the 92% of Democrats and the 42% of Republicans on some issues. However, there were not only significant differences *between* the parties in the opinions of partisans of all ages but also much bigger age differences *within* the Republican party than *within* the Democratic party. The contrast between the parties is evident in the dis-aggregation of the party totals in Table 3.2. There is virtually unanimous agreement across age groups of Democrats, but a substantial split among Republicans according to age. Whereas a majority of the Republicans in the two youngest generations agreed, less than a majority of those in the two oldest generations agreed.

An obvious implication of these *intra*-party gaps and *inter*-party gaps is that partisan patterns in Congressional and Presidential election outcomes in coming years will be among the determinants of the future of US government climate change policies.

3.4.1 China

China is not only one of the big emitting countries, it is also among the most vulnerable to the effects of the emissions. This section addresses its emissions and mitigation policies first and then its vulnerabilities and adaptation policies.

Emissions

In 2019 China's annual emissions of more than 14 gigatons of CO_2-equivalents "exceeded those of all developed countries combined," which was 27% of total global emissions (Larsen et al. 2021). Its annual emissions specifically of CO_2 increased by a factor of almost four between 2000 and 2021 (Ritchie & Roser, 2023).

The basic facts of its *cumulative* emissions, however, are quite different when put in an international comparative perspective. The sum of its CO_2 emissions from 1750 to 2021 were only 14% of the global cumulative total, whereas the US share was 24%. Among other big emitters, Germany was 5% and India was 3% (Ritchie & Roser, 2023).

In addition, because China is a major net exporter of goods, its production-based emissions have been about 10% greater than its consumption-based emissions.

Among the sources of its emissions, the coal fired electricity plants have been a major source—and a source of much contention internationally (IEA, 2022a, 2022b; Civilini & Gupta, 2023; Senlen et al. 2023a, 2023b; Vinichenko et al. 2023). A few data points are suggestive of the size, recent rate of increase, and planned capacities. In Table 3.3, the total capacities in operation as of mid-2022 are indicated. At that time China's capacity was 4.6 times greater than India's in second place, and 4.9 greater than the United States in third place. It was more than 20 times greater than each of the other seven countries in the worlds' top ten.

Plans for the future are less certain and more complex, but they reveal more striking comparisons (Senlen et al. 2023a, 2023b). During the half year following the data in Table 3.3, China's planned new coal capacity *increased* by the largest amount since 2015. Its pre-construction coal project pipeline thus became 250 gw, which was 72% of the world total world pre-construction pipeline. Meanwhile most other countries were *reducing* their planned capacities; the global *reduction* in the planned new projects was 74% since 2015.

Emission Trading System (ETS) and Other Mitigation Policies

China's ETS is the largest in the world in terms of the volume of emissions covered, though not in terms of the reductions of emissions. Pilot programs in seven cities and provinces began in 2013. In 2017 the government announced that a

Table 3.3 Top ten countries' coal capacities

Countries	Gigawatts (July 2022)[a]
China	1075
India	233
United States	218
Japan	51
South Africa	44
Indonesia	41
Russia	40
South Korea	38
Germany	38
Poland	30

Source Derived from Statistica (2023a, 2023b); also see Global Coal Plant Tracker (2023)

[a] Rounded to nearest gw

national-level program would be established, and the system was made operational in July 2021 (IEA, 2020, 2022c; Liu et al. 2022; Nakano & Kennedy, 2021; Parry et al. 2022).

The system is limited to the electricity production sector, which includes more than 2000 large firms, whose CO_2 emissions are about 40% of the country's total. A central issue is how effective the system will be in reducing emissions from coal fired plants, which have been growing significantly. In 2018, for instance, the CO_2 emissions from China's coal fired power plants were greater than those of the EU plus Japan. See IEA (2022c) for an evaluation and suggestions for improvements.

Initially, only Chinese-owned firms were included, but there was an expectation that foreign firms would be added. Also, the government has indicated that the system would expand to include other sectors such as chemicals.

Other mitigation measures, beyond the ETS, include the manufacture of electric vehicles. In mid 2023, the government extended a recently expired subsidy program when it announced a four-year program of tax exemptions equal to approximately USD 60 billion. (For a detailed analysis of the development of China's motor vehicle technology, see Sims Gallagher, 2006). China manufactures about four fifths of the world production of solar panels (Allison, 2023).

In sum, China is simultaneously the world's biggest emitter, especially from electric power plants fueled by coal, and at the same time, operator of the world's biggest emission trading system in terms of its annual volume of emissions. It is a world leader in solar panel production (Allison, 2023); Because of its large population, with exposure to cyclones in large coastal urban areas as well as inland droughts and extreme heat, it is also one of the most vulnerable countries to the impacts of climate change.

3.4.2 India

India, like China, is not only one of the big emitting countries; it is also among the most vulnerable to the effects of climate change. Like China, India has large populations exposed to coastal flooding from cyclones and population concentrations inland that are exposed to droughts, floods, and extremely high temperature events. In the first half of 2023, 14,000 people were evacuated and half a million were otherwise affected by floods, and temperatures reached 47 °C (117 °F) during a heat wave.

Extreme Heat Events

The Indian Meteorological Department (IMD) has a precise set of criteria for declaring "heat waves" and "severe heat waves" (Centre for Science and Environment [India], 2022). A "heat wave" is declared when the temperature exceeds 45 °C (113 °F), and a "severe heat wave" is declared when the temperature exceeds 47 °C (117 °F). By comparison, human body normal temperature is 35 °C (98.6 °F).

The consequences of Indian extreme heat events include (Economist, 2023):

Table 3.4 India's CO_2 emissions and CO_2-equivalent emissions of methane and nitrous oxide

GHG emissions[a]	CO_2-equivalent (kilotonnes)	
	1990	2019
Carbon dioxide (CO_2)	564	2456
Methane (CH_4)	509	657
Nitrous oxide (N_2O)	145	260
Total	1218	3373

Source Excerpted from World Bank (2022)

[a] These are three ghg emissions only, and do not include other gasses or black carbon particulate matter

- Between 2010 and 2019, annual heat-related deaths increased by 27%.
- An attribution study for a 2021 "heat pulse" found that climate change made it 30 times more likely than without climate change.
- December 2022 and February 2023 were the hottest December and February since 1901.
- It has been estimated that India loses 101 billion work hours per year because of extreme heat.
- By 2030, one estimate is that heat-related lost work hours could be 2.5–4.5% of GDP.

Emissions

India's basic emission trends and patterns can be summarized and highlighted, as in Table 3.4, for the changes over three decades from 1990 to 2019 in its CO_2 emissions and CO_2-equivalent emissions of methane and nitrous oxide. The total increased by nearly three times, and CO_2 increased by about four times.

Policies

As for sectoral shares, a central issue is the switch from fossil fuels to renewables in the energy sector. An IMF study by MacDonald and Spray (2023) has clearly stated the issue and the potential for a change:

> … combing renewable subsidies and higher tariffs on coal (roughly equivalent to ramping up India's existing excise duty on coal) would result in nearly one third lower emissions by 2030 compared to current policies. In this scenario, growing energy demand is met through a gradual increase of renewable energy and by allowing coal power to taper off, thus exceeding the goal of 50 percent non-fossil fuel electricity capacity. Under such a policy, not only would the share of renewables rise significantly but overall electricity supply would increase.

3.4.3 Comparative Ratings of Big Emitters

Table 3.5 provides summary ratings of the big national emitters featured in the chapter; the ratings are from the periodic updates by Climate Action Tracker

Table 3.5 Summary ratings of big emitters

Countries (date)	Ratings				
	Overall	Policies and actions	Conditional NDC target[a]	Unconditional NDC target[b]	Climate finance
China (6 Oct. 2023)	Highly insufficient	Insufficient	Highly insufficient	Highly insufficient	Not applicable
Germany (9 Nov. 2022)	Insufficient	Almost sufficient	Almost sufficient	Insufficient	Insufficient
India (3 Nov. 2022)	Highly insufficient	Insufficient	Critically insufficient	Insufficient	Not applicable
United States (16 Aug. 2022)	Insufficient	Insufficient	Almost sufficient	Insufficient	Critically insufficient

Source Excerpted by the author from Climate Action Tracker (2023)
[a] Conditions on actual implementations of measures—i.e. limitations—are included in the Nationally Determined Contributions (NDC) targets
[b] There are no conditions on actual implementations of measures—i.e. limitations—that are included in the Nationally Determined Contributions (NDC) targets

(2023). In addition to the overall ratings there are individual ratings for four specific dimensions, as indicted in the column headings. All of the ratings are insufficient to some degree, except that China and India are not rated for Climate Finance. The two ratings ("critically insufficient") are applied to the Indian Conditional NDC Target and the US Climate Finance category.

3.5 Vulnerable Countries

One way to identify vulnerable countries is on the basis of the frequency and severity of extreme weather events and their consequences, particular in terms of the number of deaths.

3.5.1 Extreme Weather Events

Table 3.6 identifies extreme weather events in 2021–2022, with information about the location and nature of the events, as well as their impacts on human life and economies. Not surprisingly China and India are on the list, each for more than one event; and there are also island nations of varying sizes (Fiji, Indonesia, Japan, Philippines, Sri Lanka, Maldives). The nearly 1000 deaths and well over a billion US dollar equivalents in North America and Europe are also notable. These were record-breaking years in several respects, and they were years in a new era of much greater impacts of climate change than even many climate scientists had anticipated.

Table 3.6 Major extreme weather events and their impacts (2021–2022)

Countries	Kinds of events (month)	Human deaths	Other human losses
China	Floods (July)	545	
China, Japan, Philippines	Typhoon (July)		
East Africa Ethiopia Kenya Somalia	Drought (December)		Food shortages 7.2 million 3.5 million 6 million
Fiji	Cyclone (January)		10,000 evacuated
India and Bangladesh	Cyclone (May)		1.2 million displaced
India and Nepal	Monsoon floods (October)	1153	150,000 evacuated
India, Maldives, Sri Lanka	Cyclone (May)	More than 198	
Indonesia	Cyclone		More than 22,000 displaced
Pakistan	Floods		
Philippines	Typhoon (December)	407	
South Sudan	Floods (October)		850,000 affected

Source Compiled by the author from Oxfam (2023)

The millions of lives lost or seriously disrupted in poor countries are reminders that their populations suffer disproportionately from climate change. It has been estimated, for instance, that all the countries in Africa contributes less than 4% of global emissions and that the four countries Ethiopia, Kenya, Somalia and South Sudan together contribute only 0.1%; yet, in proportionate terms they are among the world's most severely impacted.

3.5.2 Islands

Islands with low coastal areas are among the most vulnerable to the effects of climate change, especially sea level rise—and in some instances, the increased frequency of the most intensive cyclones and other storms.

Many islands are small ones that are independent countries with membership in the UN, and 38 of them belong to a formally recognized group within the UN system called the Small Island Developing States (SIDS). See Fig. 3.4 for a map of the 38 SIDS.

Table 3.7 indicates the populations and per capita GDPs of the 38 SIDS countries. The populations range from a few thousand to more than a million, and their per capita GDPs vary from a few thousand US dollar equivalents to tens of

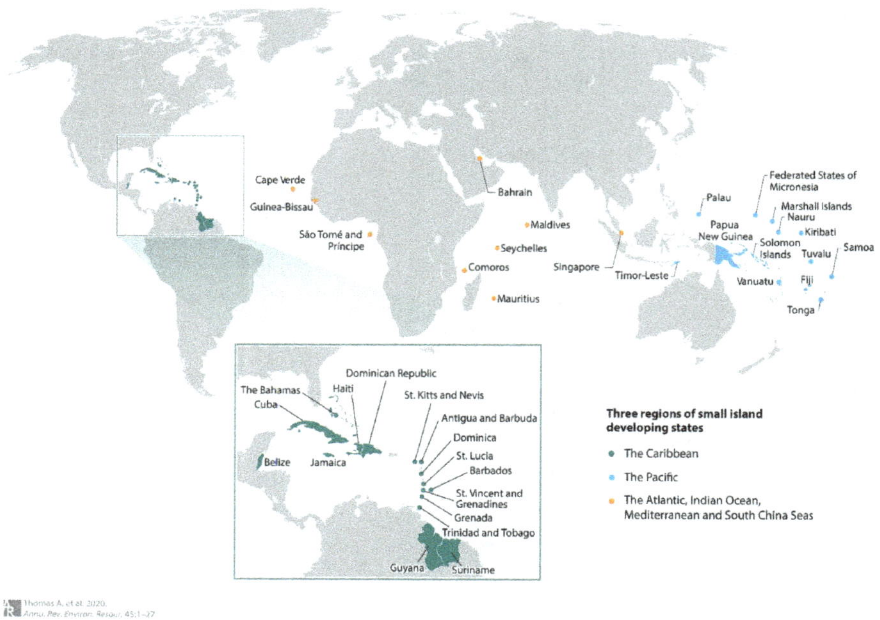

Fig. 3.4 Map of 38 Small Island Developing States (SIDS). *Source* Hickey and Unwin (2020)

thousands. (Singapore is an outlier with a per capita GDP of about US$100,000). Most of the SIDS' territorial surfaces are 3–4 m above sea level (Akiwumi, 2022). Even a one-meter sea level rise may mean that some SIDS—such as Maldives, the Marshall Islands, Kiribati and Tuvalu—will need to move many residents to higher ground or in some instances move entire populations to other countries.

See Annex 3.4 for an analysis, with specific island examples, of the kinds of vulnerabilities the SIDS face—and their prospects for the future.

There is another group of 20 islands, which are territories of other countries—or have other formalized dependencies. In the UN system, they have a special status in one of the UN regional commissions. In the Caribbean Sea, the many islands with legal status as territories—for instance, of France, the Netherlands, the UK and the US—are among those that are vulnerable to sea level rise and other climate change effects.

Table 3.7 Small Island Developing States (SIDS)

Country[a]	Region (Ocean, Sea)[b]	Population (thousands) 2020[c]	GDP per capita (USD, PPP) 2020[d]
Antigua and Barbuda	Caribbean	0.10	18,243
Bahamas	Caribbean	0.40	32,539
Barbados	Caribbean	0.29	13,350
Belize	Caribbean	0.40	6458
Cape Verde	Atlantic/Indian/So. China	20	6377
Comoros	Atlantic/Indian/So. China	870	3153
Cuba	Caribbean	11.32	
Dominica	Caribbean	0.07	10,853
Dominican Republic	Caribbean	10.95	17,936
Fiji	Pacific	0.90	13,120
Grenada	Caribbean	0.11	
Guinea-Bissau	Atlantic/Indian/So. China	2.02	1949
Guyana	Caribbean	0.79	
Haiti	Caribbean	11.54	3095
Jamaica	Caribbean	2.97	
Kiribati	Pacific	0.12	2383
Maldives	Atlantic/Indian/So. China	0.54	13,443
Marshall Islands	Pacific	0.06	4148
Mauritius	Atlantic/Indian/So. China	1272	26,840
Micronesia	Pacific	0.12	3553
Nauru	Pacific	0.01	17,820[b]
Niue	Atlantic/Indian/So. China	0.00	
Palau	Pacific	0.02	16,322
Papua New Guinea	Pacific	9.12	4287
Saint Kitts and Nevis	Caribbean	0.05	25,653
Saint Lucia	Caribbean	0.18	12,710
St Vincent and Grenadines	Caribbean	0.11	12,705

(continued)

Table 3.7 (continued)

Country[a]	Region (Ocean, Sea)[b]	Population (thousands) 2020[c]	GDP per capita (USD, PPP) 2020[d]
Samoa	Pacific	0.20	6569
Sao Tome and Principe	Atlantic/Indian/So. China	0.22	4274
Seychelles	Atlantic/Indian/So. China	98	29,470
Singapore	Atlantic/Indian/So. China	5.90	98,520[d]
Solomon Islands	Pacific	0.70	2619
Suriname	Caribbean	0.59	16,735
Timor-Leste	Pacific	1.34	4141
Tonga	Pacific	0.11	6695
Trinidad and Tobago	Caribbean	1.40	25,024
Tuvalu	Pacific	0.01	4653
Vanuatu	Pacific	0.31	3011

Sources Compiled by the author from: UN (2022), UNCTAD (2022), UNFCCC (2005, 2022): World Bank, Social Dimensions of Climate Change (2022), World Bank (2022)

[a] 38 as of 2022. The five largest—with 42.9 million—account for 66.4% of the total. They are Cuba, Dominican Republic, Haiti, Papua New Guinea, Singapore

[b] There are 16 in the Caribbean/Latin America region; 13 in the Pacific region; and 9 in the Atlantic/Indian/South China region

[c] Estimate in UN (2022)

[d] Based on PPP in World Bank (2022)

[e] Singapore is an outlier

Table 3.8 indicates the populations of the 20 islands. Several of them, of course, have relatively high-income levels; this, in combination with their affiliation with high-income countries in Europe and North America, puts them in a fundamentally different situation for undertaking adaptation measures to reduce their exposure to sea level rise and other effects of climate change. On the other hand, their populations in many cases of hundreds of thousands (or millions in the case of Puerto Rico) make them inherently vulnerable to large scale losses, especially those that are located in cyclone-prone zones.

Table 3.8 Island territories in UN Regional Commissions

Member [territory or other status of]	UN regional commission	Population (thousands) 2020
American Samoa [US]	Asia/Pacific	55
Anguilla [UK]	Asia/Pacific	15±
Aruba [Netherlands]	Latin America/Caribbean	117
Bermuda [UK]	Latin America/Caribbean	63.9
British Virgin Islands [UK]	Latin America/Caribbean	30.0
Cayman Islands [UK]	Latin America/Caribbean	66
Cook Islands [New Zealand, "free association"]	Asia/Pacific	17±
Curacao [Netherlands]	Latin America/Caribbean	165
French Polynesia [France]	Asia/Pacific	281±
Guadeloupe [France, Department]	Latin America/Caribbean	396±
Guam [US]	Asia/Pacific	316
Martinique [France, Department]	Latin America/Caribbean	376±
Montserrat [UK]	Latin America/Caribbean	492
New Caledonia [France, "collective special"]	Asia/Pacific	540
Niue [New Zealand, "free association"]	Asia/Pacific	2±
Northern Mariana Islands [US]	Asia/Pacific	580
Puerto Rico [US]	Latin America/Caribbean	3194
Sint Maarten [Netherlands][b]	Latin America/Caribbean	41±
Turks and Caicos Islands [39]	Latin America/Caribbean	39
US Virgin Islands [US]	Latin America/Caribbean	87±

Sources UN Department of Economic and Social Affairs (2022), UN Sustainable Development Knowledge Platform (2022), World Bank (2022)

[a] Other than UN members listed in Table 4.1

[b] Does not include French part of the island

3.6 Multiple Crises in East Africa

Although islands present special profiles of vulnerability to the impacts of climate change, there are many other countries that have already been experiencing catastrophic impacts. Countries in East Africa are among the most seriously impacted. As Box 3.4 indicates, in addition to recurrent severe droughts, the countries have experienced multiple shocks that exacerbate the effects of the droughts on food security. The countries thus represent instructive examples of how droughts caused by climate change can interact with other crises to produce extraordinarily catastrophic circumstances.

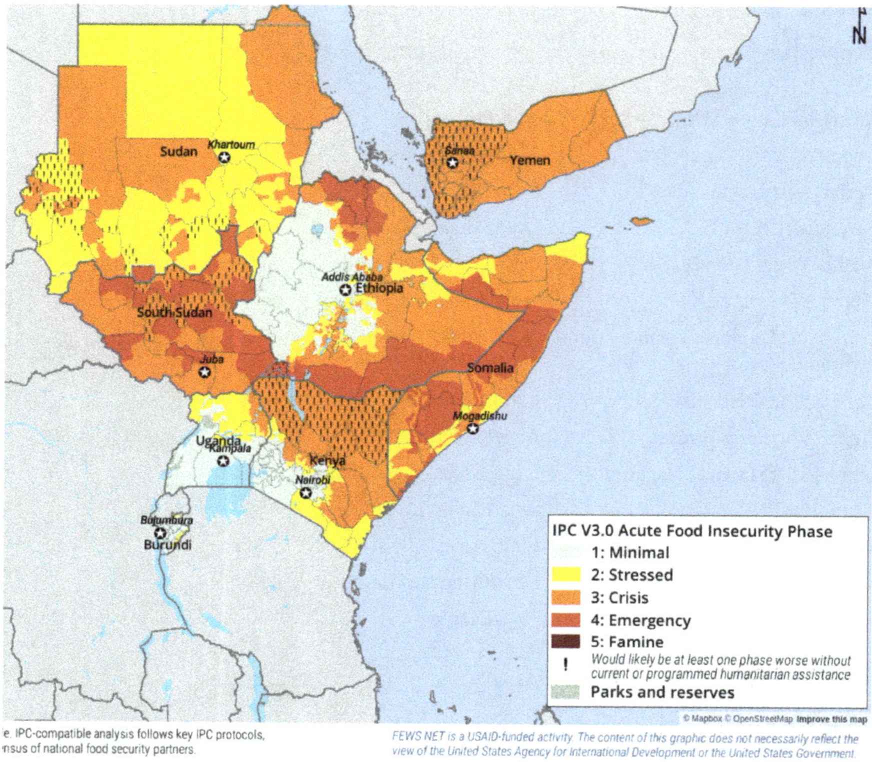

Fig. 3.5 Map of food insecurity in East Africa (Ethiopia, Kenya, Somalia, Southern Sudan). *Source* US NASA (2022)

The analysis in the box covers a decade beginning in 2011. The accompanying map in Fig. 3.5 reveals the substantial variations within each of the four countries in the severity of the impact on food security.

Box 3.4: Recurrent Catastrophic Droughts in East Africa (Ethiopia, Kenya, Somalia, South Sudan)

Across the Horn of Africa, households now face multiple concurrent shocks to food security. The ongoing drought occurs in a volatile context where conflict, insecurity, economic challenges, and desert locusts are also straining rural livelihoods. … [T]he multiple shocks they face have pushed many to a breaking point, where their ability to further cope is now almost exhausted. …

Drought is among the most devastating of natural hazards—crippling food production, depleting pastures, disrupting markets and, at its most extreme, causing widespread human and animal deaths. Droughts can also lead to increased migration from rural to urban areas, placing additional pressures on declining food production. Herders are often forced to seek alternative sources of food and water for their animals, which can create conflict between communities, competing for the scarcely available resources. ...

In 2011, the drought considered to be "the worst in 60 years" at that time, combined with serious access issues, pushed Somalia into famine. Up to 260 000 people—half of them children—died, and the drought caused massive displacement across the region. The response was deemed to be too little too late. ...

Between 2016 and 2019, the region faced six out of seven below-average rainy seasons. The worst was avoided thanks to anticipatory action, including the use of crisis modifier modalities, rapidly mobilized additional resources in 2016 and sustained large-scale humanitarian assistance throughout the period of concern....

Since October 2020, the region has entered into a new episode of worsening conditions [2022], for the third consecutive season....

...[T]he preliminary results of the Kenya mid-season assessment October–December 2021 found that over 1.4 million livestock ... died over the [previous], three months. Similarly, the preliminary results of the Kenya mid-season assessment October–December 2021 show[ed] that maize production [would] be up to 70% below average in marginal agricultural areas....

Source: Food and Agriculture Organization (FAO) (2022).

3.7 Compensation Issues: "Loss and Damage"

For more than 30 years, low-income countries have been calling upon high-income countries to provide financial compensation for "loss and damage" from the impacts of climate change. This could be in addition to any financial aid for adaptation projects that are specifically designed to reduce the effects of sea level rise, hurricanes and other storms, as well as the effects of floods and periods of extreme heat that are threats to human life, property and agricultural production. Of course, there are also long-standing international economic development programs of many high-income countries in Europe, North America, East Asia and the South Pacific. Whether these would be credited toward loss and damage compensation payments is an issue.

COP 15 Agreement in 2009

At the COP 15 in Copenhagen in 2009, there were new climate-specific financial assistance goals and policies developed and announced. An important one—at least symbolically—was the agreement that the developed countries would contribute US$ 100 billion per year starting in 2020 and continuing through 2025. In any case, discussion of this goal was eclipsed by yet more announcements and agreements in different international institutional contexts. At a G7 meeting in mid-2022, a US$ 600 billion plan was announced as a counter to China's Belt and Road Initiative (Lemire & Mathiesen, 2022). All these initiatives have in common large amounts of money pledged over long periods of time; the amounts of actual transfers are of course another matter, as documented below.

Over the several decades when the issue has been on the agendas of many COPs and other venues, discussions of the issue have been marked by two themes. One is that low-income countries, most of which are also low emitters of greenhouse gasses on a per capita basis, contend that the losses and damages that they incur are the result of the emissions over the centuries since the beginning of the industrial revolution in the mid-eighteenth century until the present. Indeed, the high-income countries' share of emissions over that time period is large. It is the *cumulative level* of atmospheric gasses and aerosols that causes the warming, with net *annual rates* making incremental additions to the cumulative level. The high-income countries, it is argued, therefore have a moral responsibility to compensate for this historically demonstrable injustice. On the other hand, the US and some other high-income countries point to China and India for their high and increasing current rates of emissions as part of the problem; they are also concerned that paying compensation could be used as an implicit recognition of responsibility and thus the basis for a large number of legal liability cases against governments and corporations in high-income countries.

For decades, there was an impasse about what to do. Then, in September 2022, as pressure from vulnerable countries was increasing ahead of the COP 27 meeting to establish a special international "loss and damage" fund to assist pour countries with their adaptation expenses, Denmark announced its support for the establishment of such a fund. At the same time, it announced a contribution of DK 100 million (equivalent to more than US$ 13 million) to aid—among other regions—the semi-arid Sahel region at the southern edge of the Sahara desert in northwestern Africa which includes portions of Burkina Faso, Chad, Mali, Mauritania, Niger, Nigeria, Senegal, and Sudan.

The issues are complicated by the existence of six well established channels of financial assistance to developing countries:

- Bilateral Official Development Assistance (ODA) from countries that are members of the Development Assistance Committee (DAC) of the Organization for Economic Cooperation and Development (OECD).
- Other Official Flows (OOFs) from DAC members—which are non-concessional or non-developmental flows.
- Bilateral flows from non-DAC countries, including concessional, non-concessional and non-developmental.
- Multilateral institutions, including UN agencies.
- Multilateral Development Banks (MDBs), including the World Bank, regional development banks, and other similar institutions.
- Philanthropic foundations' concessional flows.

Actual Transfers

The amounts according to the providers, the recipients, and the projects are tracked by the OECD (2022a, 2022b) (Table 3.9).

The table reflects both the persistence of some basic patterns and some changes over time, particularly during the covid shock of 2018, 2019 and 2020. Public sources were much greater than private sources—with the former increasing from three-fourths of the total in 2013 to five-sixths in 2020. The trend was evident during the covid shock: a slight increase in the absolute amount from public sources and a slight decrease from private sources. The overall total increased by only 4% from 2018 to 2020.

A proposal by Avinash Persaud, a finance official of Barbados, is based on a fundamentally different approach to the challenges of increasing investments in developing countries for adapatation as well as mitigation—an approach he claims could lead to transfers of more than a trillion USD per year (UN (2023)).

Table 3.9 International financial flows from developed to developing countries for climate action

Sources[a]	US Dollars (billions)[b]				Percent of total[c]
	2013	2018	2019	2020	2020
Bilateral public finance (1)	22.5	32.0	28.7	31.4	37
Multilateral public finance attributable to developed countries (2)	15.5	30.5	34.7	36.9	44
Multilateral development banks	13.0	26.7	30.5	33.2	40
Multilateral climate funds	2.2	3.5	3.8	3.5	4
Inflows to multilateral institutions (where outflows unavailable)	0.3	0.3	0.3	0.2	< 1
Subtotal: Items 1 + 2 above	*38.0*	*62.1*	*63.4*	*68.3*	*82*
Climate-related officially-supported export credits (3)	1.6	2.7	2.6	1.9	2
Subtotal: Items 1 + 2 + 3 above	*39.5*	*64.8*	*66*	*70.2*	*84*
Mobilized private climate finance (4)	12.8	14.7	14.4	13.1	16
By bilateral public finance	6.5	3.8	5.8	5.1	6
By multilateral public finance attributable to developed countries (2)	6.2	11.0	8.6	8.0	10
Total: Items 1 + 2 + 3 + 4[d]	*52.4*	*79.9*	*80.4*	*83.3*	*100*

Source Excerpted by the author from OECD (2022a); 2020 percents computed by the author. Also see OECD (2021, 2022b, 2023), Oxfam (2023), Elliot (2023), Harvey (2021)

[a] These are only flows from developed economies to developing economies. They do not include flows among developing countries or internal domestic flows within those countries

[b] Current, not constant, USD

[c] Computed by the author; not in original source

[d] Totals and subtotals may reflect rounding errors

These issues about international financial transfers from rich to pour countries inevitably entail ethical issues. However, there are many other ethical issues, as discussed in the next section.

3.8 Ethical Issues

This chapter includes a broad range of ethical issues associated with greenhouse gas and particulate matter emissions. Ethical issues inherently involve questions of values, including which actions are ethically acceptable and unacceptable, on the basis of what values. Because ethical perspectives are affected by cultural differences as well as historical experiences, it is often difficult to reach a consensus on any given ethical question. Nevertheless, in the case of global climate change issues, at an international level of analysis adopted in this chapter in the above sections, there is an obvious question about whether it is fair that rich countries are major emitters and that poor countries are the most vulnerable to the consequences of the emissions.

That issue is the obvious the focus in this chapter. In particular, there are familiar international economic and political concerns at the core of contemporary international economic development issues. The core of the conflict between the high-income and the low-income countries is that the former's emissions have created a global problem as they became increasingly wealthy over the decades, while the latter are disproportionately suffering the economic, health and other consequences of the emissions. Although the issues emerged many years ago, even before COP 15 in 2009, they became much more salient in 2022 at COP 27. Thus, now in the second decade of the twenty-first century, they are front and center on the international climate change agenda.

In addition to this international level of analysis, however, the analysis in the chapter also notes that there are differences among groups within countries. At this sub-national level of analysis, there is an analogous question about whether it is fair that relatively wealthy groups are disproportionate emitters and relatively poor groups within the same national society are disproportionately vulnerable to the consequences of the emissions.

Finally, there are questions at the individual level of analysis that focus on the responsibilities of individuals to take some actions and avoid other actions. This is a universally applicable issue in all countries and groups within them. It is sometimes at the core of issues about fairness across generations as well.

In sum, at all levels of analysis, the key ethical question can be succinctly stated as: "Who owes how much to whom for what?" Perhaps that is a question for each reader to address on the basis of her/his own ethical perspectives.

3.9 Conclusion

As the current impacts of climate change emissions have become more tangible and salient—and as future projections continue to anticipate more frequent and widespread impacts around the world—the most vulnerable people and places have been receiving more attention by policymakers and publics. The basic features of vulnerable people and places are well known: people in low-lying coastal areas in many countries of all sizes are vulnerable to the damages from rising sea levels, more damaging storm surges, the most serious cyclones, and floods. People in agricultural areas and near forests are especially vulnerable to the combination of droughts and wildfires. In these and other situations, low-income people are especially vulnerable because of the economic costs of recovery from such events.

Actual extreme weather events enable public health specialists, insurance company adjusters, economists, accountants, government agency officials and others to estimate the economic costs. A single extreme weather event can cause damages totaling tens-to-hundreds of billions of US dollar equivalents; a single extreme event can also cause hundreds-to-thousands of human deaths. Increasing research activity in "attribution studies" is providing evidence about the contributions of climate change to individual extreme weather events.

In short, much is known about the causes of climate change and the effects on vulnerable people and places. As for conflicts over financial assistance for adaptation projects and/or compensation for loss and damages, those are issues for diplomats, legislators, legal experts, government officials, and ethicists to ponder.

Annex 3.1: EU Emission Trading System (ETS)

The EU ETS includes all of the EU-27 countries plus three other countries—Iceland, Lichtenstein and Norway. After it was legally created in 2003, it began operations in 2005; during phase 1, it conducted tests on a pilot basis.

The key design issues were:

(1) the amount of the annual cap on the creation of EU Allowances (EUAs)—in total, for national members and for installations;
(2) what industries to include;
(3) what kinds of emissions to include;
(4) how many free allocations, if any;
(5) whether and how to impose limitations on EUA prices;
(6) administrative systems for monitoring, reporting and verifying (MRV) installations' emissions;
(7) enforcement procedures and penalties for non-compliance;
(8) what countries to include.

The decisions for these issues evolved over time in an iterative learning process organized into phases of varying length from three to eight years. Thus, after phase 1, each phase has been marked by a series of revisions. The resulting key features of each phase, have involved the following:

Phase 1 (2005–2007)—Pilot testing phase. Members' national plans set total and installation allocations. Members could auction 10% of allowances. CO_2 emissions covered power and heat generation, iron, steel, cement, oil refineries. Non-compliance penalties were 40 euros per tonne.

Phase 2 (2008–2012)—Total EU cap set at 6.5% less than 2005 emissions. Nitrous oxide added to coverage. Aviation flights within ETS area included, beginning in 2012. Three non-EU countries—Iceland, Lichtenstein and Norway—added to ETS. Non-compliance penalties were increased to 100 euros per tonne. Because of new credits that could be applied from the UNFCCC Kyoto Protocol Clean Development Mechanism (CDM) and Joint Implementation (JI) program—and the global financial crisis of 2008—the price of an EUA declined from 30 to less than 7 euros.

Phase 3 (2013–2020)—Many revisions. Cap reduced by 1.74% per year to meet 21% reduction target by 2020 compared with 2005. Increase in auctioned EUAs and decrease in free EUAs. EU ETS Auctioning Regulation applied to make auctions more transparent. In the first year of the new auctioning process, more than 40% of verified EUAs were auctioned. New product-specific benchmarks applied in several industries, with declining amounts over time—except 100% for industries exposed to international carbon leakages. EUA price declined to 3–7 euros because of surplus phase 2 EUAs transferred to phase 3. Scheduled auction of 900 million EUAs postponed to the end of phase 3.

Phase 4 (2021–2028)—For sectors covered by the ETS, establish a ghg emissions reduction target for 2030 at 43% compared with 2005. Sectors most exposed to international carbon leakage receive 100% free allocations; free allocations for less exposed sectors phased out after 2026 from a maximum of 30–0%. Set aside free allowances for new and growing installations. More flexible rules and annual adjustments for better alignment of allocations with installations' actual production levels. Update benchmarks for establishing free allocation amounts to reflect windfall profits and technological progress since 2008. Establish an Innovation Fund to support innovative technologies at a level at least the equivalent of 450 million allowances at market value. Establish a Modernisation Fund for the power sector and other energy systems for greater efficiency and a just transition in carbon-dependent regions in ten low-income member countries. Beginning in 2023, limit the number of allowances in the Market Stability Reserve (MSR) to the auction volume of the previous year.

Sources: Böhringer and Lange (2013), Climate Policy Info Hub (2022), European Commission (2022).

Annex 3.2: Emissions and Mitigation Policies of the UK

Because the industrial revolution of the eighteenth and nineteenth centuries began in the UK—with coal as the principal fuel—that country has continued to be a leader in its total cumulative ghg emissions. In the late 20th and early twenty-first centuries, however, government policies changed as the results of climate science studies and the statements of international conferences clarified the nature, sources and effects of greenhouse gas emissions. Indeed, a speech by Prime Minister Thatcher at the UN General Assembly—noted in a header quote of this chapter—publicly sounded the alarm. (Some years later, however, she was opposed to many proposals and policies to reduce ghg emissions.) In subsequent years, there was agreement among conservative party and labour party leaders that climate change was a serious problem that needed serious attention. That consensus was interrupted, however, in 2022 when the Truss government showed more interest in developing fossil energy sources and less interest in addressing climate change issues. She appointed a well-known climate change denier Conservative MP Rees-Mogg—a long-time opponent of action on climate change—to be Minister of Business, Energy and Industrial Strategy.

Annex 3.3: Comparisons of Consumption-Based and Production-Based Accounting of National Emissions

"Consumption-based accounting" (CBA)—which is also called "carbon footprint accounting"—indicates the emissions of a country including the carbon content of imports but excluding the carbon content of exports. "Production-based accounting" (PBA)—which is also called "territorial accounting"—ignores imports and exports and thus reflects a "domestic" focus. The latter is used for UNFCCC emissions reporting and is also the basis for most other emissions data bases, including those of the International Energy Agency (IEA) and the EU. There are several sources of comparative data that are especially useful for understanding the significance of the differences in the indicators: UK Office of National Statistics (2019) for the UK and other countries; Eora, which is a world-wide multi-region international trade input–output data set (Worldmiro, 2022).

The change in the UK over a period of more than 4 decades (1970–2015) was substantial, as noted in Table 3.10. Part of the story, as noted in the body of the chapter, was a shift from domestically-produced coal to imported coal.

The large changes over time as well as the large differences in countries' profiles as net exporters or net importers of carbon are evident in Table 3.11. The absolute magnitudes of the increases for China as a leading net exporter and the US as a leading net importer by 2015 are both obvious.

Table 3.10 Comparisons of UK's consumption and territorial emissions levels (1970, 2015)

	1970	2015
Consumption-based (million tonnes)	700.9	656.6
Territory-based (million tonnes)	685.0	402.5
Difference C > T (million tonnes)	15.9	254.1
Difference C > T (%)	2	63

Source Compiled by the author from UK Office of National Statistics (2019: Fig. 9)

Table 3.11 Carbon content of six big emitters' net exports or imports (mil. tonnes, 1992, 2015)

	1992	2015
Net exporters		
China	364	1481
Russia	276	408
India	42	256
Net importers		
US	43	717
UK	97	258
Japan	227	218

Source Compiled by the author from UK Office of National Statistics (2019: Fig. 10)

Annex 3.4: Vulnerabilities of SIDS

In spite of variations in geographical, physical, climatic, social, political, cultural and ethnic features, and economic development, small island developing States share certain characteristics that underscore their overall vulnerability to climate change:

- Generally limited natural resources, with many already heavily stressed from unsustainable human activities.
- A concentration of population, socio-economic activities, and infrastructure along the coastal zone.
- High susceptibility to frequent and more intense tropical cyclones (hurricanes) and to associated storm surge, droughts, tsunamis and volcanic eruptions.
- Dependence on water resources for freshwater supply that are highly sensitive to sea-level changes.
- Relative isolation and great distance to major markets, affecting competitiveness in trade.
- Extreme openness of small economies and high sensitivity to external shocks.
- Generally high population densities and in some cases high population growth rates.
- Inadequate infrastructure in most sectors.

- Limited physical size, effectively eliminating some adaptation options to climate change and sea-level rise.
- Insufficient financial, technical and institutional capacities, seriously limiting the capacity of SIDS to mitigate and adapt to any adverse impacts of climate change.
- A few examples of the vulnerabilities in specific islands illustrate well this high level of susceptibility to the adverse impacts of climate change:
- In Barbados and many other islands, almost all foods, fuels, construction materials and other goods are imported.
- In the Maldives and Papua New Guinea, some 50–80% of the land area is less than 1 m above mean sea level.
- In the Seychelles, about 80% of the infrastructure and population are found along the coast.
- Critically limited resources are required to address pressing short-term environmental problems in Grenada.
- In Palau, prolonged droughts are experienced during El Niño Southern Oscillation events.
- Throughout the South Pacific region, frequent and more intense tropical cyclones, as well as climate-related and other extreme events, were experienced during the 1990s.

Due to the geographic location of the SIDS and the profound influences of oceanic circulation systems, natural precipitation varies from one year to the next much more than in other countries. This can lead to various forms of extreme rainfall events, such as droughts and floods, that have a wide range of adverse impacts—including some catastrophic damages—on natural and human systems.

Climate projections suggest that significant climate change and sea-level rise are expected in all regions during the twenty-first century. Increases in atmospheric concentrations of GHGs due to human activities over the past 100 years will continue to alter the climate and related systems on Earth in the coming century, if not for longer. Subsequently, SIDS face the prospect of increased challenges to their efforts to achieve sustainable development.

Indeed, an ensemble of climate model simulations for seasonal temperature and rainfall changes in the four regions by the 2050s and 2080s project the following changes:

- Temperature increase is projected for all regions and for both seasons.
- Warming over the Mediterranean and the Caribbean Sea areas is higher during northern hemisphere winters for both time periods, whereas warming in the other two regions exhibits different seasonal patterns for the 2050s and 2080s.
- For the 2080s, SIDS in the Mediterranean area are projected to experience the highest warming, with surface air temperature rising by 3.9 °C for December–February and 4.5 °C for June–August.
- Projections show a dominantly increased pattern in seasonal rainfall for the four regions, with islands in the Mediterranean area getting the most increase

in rainfall during northern hemisphere winters (by the 2080s, 16% higher than the 1961–1990 average level).

- The largest decline in seasonal rainfall is projected for SIDS in the Caribbean area, with a reduction during northern hemisphere summers of nearly 20%.

These projected changes are likely to exacerbate the current climate-related stresses in various SIDS. Higher temperatures are expected to adversely affect the health of some island inhabitants who already suffer through heat waves and associated increased outbreaks of vector-borne diseases. The health of important marine species such as coral reefs will also suffer.

Changes in seasonal rainfall patterns may take the form of more frequent and more intense droughts and floods for many of the already troubled SIDS. Vulnerability and adaptation assessments undertaken by SIDS also support the prediction that sea level will rise worldwide as a consequence of climate change. This is consistent with the conclusions of the Third Assessment Report of the Intergovernmental Panel on Climate Change, which indicated that by the end of the twenty-first century global sea level will rise 9–88 cm depending on regional variations in ocean circulation and other geophysical conditions. Sea-level rise will vary across the planet, with some regions (such as the Pacific) experiencing a greater increase than the global average.

Although natural systems and people living in SIDS have developed a wide range of adaptive strategies and measures to cope with a certain degree of change in climate and sea level, in many of these nations the environment and various biological systems are already close to their tolerance limits or are experiencing climate-related stress. Their vulnerability to climate change—already high—thus worsens with each passing year.

Source: UNFCCC (2005).

Questions to Ponder

1. Which countries are both among the big emitters and among the most vulnerable to the socio-economic effects of emissions?
2. What kind of formula could be developed to determine the amounts of money to be contributed to an international climate change compensation fund?
3. Do you think only recent years' emission amounts or cumulative long-term amounts should be the basis for determining countries' contributions to an international compensation fund?
4. Is it true that the world's biggest ghg emitter is China or the US or the UK—depending on the indicator used? Why?
5. What are the net carbon export and import profiles for the six big emitting countries for the latest year of data available? Include both absolute levels and levels adjusted for the countries' GDPs.

6. In the relationship between the developed economy countries that are big emitters and the developing economy countries that are vulnerable to the effects of the emissions: Who owes how much to whom for what?

References

Akiwumi, P. (2022). UNCTAD, Prosperity for All. Climate finance for SIDS is shockingly low: Why this needs to change. https://unctad.org/news/climate-finance-sids-shockingly-low-why-needs-change. Accessed June 11, 2022.

Allison, G. (2023) China's dominance of solar poses difficult choices for the west. *Financial Times,* 22 June. https://www.ft.com/content/fd8e7175-9423-4042-a6f7-c404afdfcda4. Accessed June 22, 2023.

Böhringer, C., & Lange, A. (2013). Design of the EU emissions trading system. In *Encyclopedia of energy, natural resource, and environmental economics*. https://www.sciencedirect.com/topics/economics-econometrics-and-finance/european-union-emission-trading-system. Accessed June 2, 2022.

Brewer, T. (2022). *Transforming US Climate Change Policies: 2021 and Beyond*. Policy Brief. Springer. https://link.springer.com/book/9783030997175.

Centre for Science and Environment [India]. (2022). Heat waves in India. https://www.cseindia.org/. Accessed June 24, 2023.

Civilini, M., & Gupta, A. (2023). Chinese coal boom a 'direct threat' to 1.5C goal, analysts warn. https://climatechangenews.com/2023/03/14/chinese-coal-boom-a-direct-threat-to-1-5c-goal-analysts-warn/. Accessed June 22, 2023.

Climate Action Tracker. (2023). Climate tracker countries. https://climateactiontracker.org/countries/. Accessed June 23, 2023.

Climate Policy Info Hub. (2022). The EU emissions trading system: An introduction. https://climatepolicyinfohub.eu/eu-emissions-trading-system-introduction. Accessed June 2, 2022.

Comer, B. & Osipova, P. (2021). Accounting for well-to-wake carbon dioxide equivalent emissions in maritime transportation climate policies. https://theicct.org/wp-content/uploads/2021/06/Well-to-wake-co2-mar2021-2.pdf. Accessed June 12, 2022.

Economist. (2023). Large, creative AI models will transform lives and labour markets. *The Economist*. https://www.economist.com/interactive/science-and-technology/2023/04/22/large-creative-ai-models-will-transform-how-we-live-and-work. Accessed May 7, 2023.

Elliot, L. (2023). The west has broken its promises to developing countries—And we're all paying the price. *The Guardian,* 15 June. https://www.theguardian.com/commentisfree/2023/jun/15/west-promises-developing-countries-summit-paris-climate-vaccines-budgets. Accessed June 21, 2023.

European Commission. (2019a). The European green deal. https://eur-lex.europa.eu/legal-content/EN/TXT/?qid=1576150542719&uri=COMpercent3A2019percent3A640percent3AFIN. Accessed May 31, 2022.

European Commission. (2019b). What is the European green deal? file:///C:/Users/Thomas percent20Brewer/Downloads/What_is_the_European_Green_Deal_en.pdf.pdf. Accessed May 31, 2022.

European Commission. (2021a). Fit for 55. Accessed June 1, 2022.

European Commission. (2021b). European green deal: Commission proposes transformation of EU economy and society to meet climate ambitions. https://ec.europa.eu/commission/presscorner/detail/en/IP_21_3541. Accessed June 1, 2022.

European Commission. (2021c). Delivering the European green deal. https://ec.europa.eu/clima/news-your-voice/news/delivering-european-green-deal-2021-07-14_en. Accessed June 1, 2022.

European Commission. (2022). Climate action: EU emissions trading system (EU ETS). https://ec.europa.eu/clima/eu-action/eu-emissions-trading-system-eu-ets_en. Accessed June 2, 2022.

European Environment Agency. (EEA). (2022). Annual European Union greenhouse gas inventory 1990–2020 and inventory report: Submission to the UNFCCC Secretariat. https://www.eea.europa.eu/publications/annual-european-union-greenhouse-gas-1. Accessed July 1, 2022.

European Parliament. (2022). Legislative train schedule: Fit for 55 package under the European green deal. https://www.europarl.europa.eu/legislative-train/package-fit-for-55. Accessed June 1, 2022.

European Union, Official Journal. (2021). Regulation (EU) 2021/1119 of The European Parliament and of The Council. https://perma.cc/2GM3-9W99. Accessed June 1, 2022.

Eurostat. (2022). Greenhouse gas emission statistics - emission inventories. https://ec.europa.eu/eurostat/statistics-explained/index.php? title=Greenhouse_gas_emission_statistics_-_emission_inventories. Accessed June 15, 2023.

Food and Agriculture Organization. (FAO). (2022) Drought in the Horn of Africa. Rapid response and mitigation plan to avert a humanitarian catastrophe. January–June 2022. https://www.fao.org/3/cb8280en/cb8280en.pdf. Accessed June 23, 2022.

Global Coal Plant Tracker. (2023). Coal-fired power capacity in China. https://docs.google.com/spreadsheets/d/1sHBsK_Ez7C9XA4HKRQSvopO4IvGSLz65jxdG0GQXeVk/edit#gid=1236850196. Accessed June 22, 2023.

Hansen, J. (1988). The greenhouse effect: Impacts on current global temperature and regional heat waves; Presented to United States Senate Committee on Energy and Natural Resources, June 23, 1988. https://pulitzercenter.org/sites/default/files/june_23_1988_senate_hearing_1.pdf. Accessed February 16, 2022.

Harvey, F. (2021). Rich countries not providing poor with pledged climate finance, analysis says. https://www.theguardian.com/environment/2021/sep/20/rich-countries-not-providing-poor-with-pledged-climate-finance-analysis-says. Accessed June 21, 2023.

Hickey, G. & Unwin, N. (2020). Addressing the triple burden of malnutrition in the time of COVID-19 and climate change in Small Island Developing States: what role for improved local food production? Food Security, 12(2): 1-5. https://www.researchgate.net/publication/342804854_Addressing_the_triple_burden_of_malnutrition_in_the_time_of_COVID-19_and_climate_change_in_Small_Island_Developing_States_what_role_for_improved_local_food_production. Accessed June 15, 2023.

International Energy Agency (IEA). (2020). China's emission trading scheme. https://www.iea.org/reports/chinas-emissions-trading-scheme. Accessed June 23, 2023.

International Energy Agency (IEA). (2022a). Coal in net zero transitions. https://www.iea.org/reports/coal-in-net-zero-transitions. Accessed January 12, 2023.

International Energy Agency (IEA). (2022b). Coal 2022. https://www.iea.org/reports/coal-2022. Accessed January 12, 2023.

International Energy Agency (IEA). (2022c). Enhancing China's ETS for carbon neutrality. https://www.iea.org/reports/enhancing-chinas-ets-for-carbon-neutrality-focus-on-power-sector. Accessed June 23, 2023.

Larsen, L., et al. (2021). China's greenhouse gas emissions exceeded the developed world for the first time in 2019. Rhodium Group. https://rhg.com/research/chinas-emissions-surpass-developed-countries/. Accessed June 22, 2023.

Lemire, J., & Mathiesen, K. (2022). G7 unveils $600B plan to combat China's belt and road. *Politico*, 26 June. https://www.politico.eu/article/g7-unveils-600b-plan-to-combat-chinas-belt-and-road/?itm_source=parsely-api?itm_campaign=parsely_recommended_widget-4&itmMedium=site_widget&itmSource=parsely_recommended_widget&itm_content=widget_item-2. Accessed June 27, 2022.

Liu, J., et al. (2022). China's national ETS. *Science Direct, Energy Reports, 8*, 428–437. https://www.sciencedirect.com/science/article/pii/S2352484722006564. Accessed June 23, 2023.

MacDonald, M., & Spray, J. (2023). *India can balance curbing emissions and economic growth.* International Monetary Fund. https://www.imf.org/en/News/Articles/2023/03/06/cf-india-can-balance-curbing-emissions-and-economic-growth. Accessed June 24, 2023.

Nakano, J., & Kennedy, S. (2021). Critical questions. China's New National Carbon Trading Market. Center for Strategic and International. https://www.csis.org/analysis/chinas-new-national-carbon-trading-market-between-promise-and-pessimism. Accessed June 23, 2023.

OECD. (2021). Climate finance provided and mobilised by developed countries. https://www.oecd.org/env/climate-finance-provided-and-mobilised-by-developed-countries-aggregate-trends-updated-with-2019-data-03590fb7-en.htm. Accessed June 21, 2023.

OECD. (2022a). Aggregate trends of climate finance provided and mobilised by developed countries in 2013–2020. Climate Finance and the USD 100 Billion Goal. https://www.oecd.org/climate-change/finance-usd-100-billion-goal/. Accessed June 21, 2023.

OECD. (2022b). Forum on green finance and investment. https://www.oecd-events.org/oecd-forum-on-green-finance-and-investment-2022. Accessed June 21, 2023.

OECD. (2023). OECD guidance on transition finance: ensuring credibility of corporate climate transition plans. https://www.oecd-ilibrary.org/sites/7c68a1ee-en/1/3/1/index.html?itemId=/content/publication/7c68a1ee-en&_csp_=de7026e6bbb9a2098a2b3b13291bc473&itemIGO=oecd&itemContentType=book. Accessed June 21, 2023.

Oxfam. (2023). Climate finance shadow report 2023: Assessing the delivery of the $100 billion commitment. https://policy-practice.oxfam.org/resources/climate-finance-shadow-report-2023-621500/. Accessed June 20, 2023.

Parry, I., Black, S., & Zhunussova, K. (2022). Carbon taxes or emissions trading systems? Instrument choice and design. IMF Staff Climate Note 2022/006. International Monetary Fund (IMF). https://www.imf.org/en/Publications/staff-climate-notes/Issues/2022/07/14/Carbon-Taxes-or-Emissions-Trading-Systems-Instrument-Choice-and-Design-519101. Accessed 23 June 2023.

Pew. (2021). Key findings: How Americans' attitudes about climate change differ by generation, party and other factors. https://www.pewresearch.org/short-reads/2021/05/26/key-findings-how-americans-attitudes-about-climate-change-differ-by-generation-party-and-other-factors/. Accessed June 20, 2023.

Ritchie, H., & Roser, M. (2023). China. Our world in Data. https://ourworldindata.org/co2/country/china. Accessed June 22, 2023.

Senlen, O., et al. (2023a). Diverging pathways: China's new coal boom takes it on a detour. E3G. https://www.e3g.org/news/china-s-new-coal-boom-takes-it-on-a-detour-while-rest-of-world-drives-forward/. Accessed June 22, 2023.

Senlen, O., et al. (2023b). Driving forward: World outside china closes in on "no new coal." E3G. https://www.e3g.org/publications/world-drives-forward-on-no-new-coal-but-china-takes-a-detour/. Accessed June 22, 2023.

Sims Gallagher, K. (2006). *China shifts gears.* MIT.

Statistica. (2023a). Countries with largest installed capacity of coal power plants worldwide as of July 2022. https://www.statista.com/statistics/530569/installed-capacity-of-coal-power-plants-in-selected-countries/. Accessed June 22, 2023.

Statistica (2023b). Germany. https://www.statista.com/. Accessed June 22, 2023.

Thatcher, M. (1989). Speech to United Nations General Assembly, 8 November 1989. https://www.margaretthatcher.org/document/107817. Accessed July 7, 2021.

UK Office of National Statistics. (2019). The decoupling of economic growth from carbon emissions: UK evidence, section 7. International trade of carbon emissions. https://www.ons.gov.uk/economy/nationalaccounts/uksectoraccounts/compendium/economicreview/october2019/thedecouplingofeconomicgrowthfromcarbonemissionsukevidence. Accessed May 31, 2022.

United Nations Framework Convention on Climate Change (UNFCCC). (2005). Climate change, small island developing states. https://unfccc.int/resource/docs/publications/cc_sids.pdf. Accessed June 10, 2022.

United Nations Conference on Trade and Development (UNCTAD). (2022). UN list of least developed countries. https://unctad.org/topic/least-developed-countries/list. Accessed June 11, 2022.

United Nations Framework Convention on Climate Change (UNFCCC). (2022). Small island developing states needs survey: Quantitative and qualitative. https://unfccc.int/sites/default/files/resource/Needspercent20Surveypercent20ofpercent20SIDSpercent20-percent20Final.pdf. Accessed June 10, 2022.

United Nations Department of Economic and Social Affairs. (2022). *Population dynamics*. https://population.un.org/wpp/Download/Standard/Population/. Accessed June 18, 2022.

United Nations Sustainable Development Knowledge Platform. (2022). Non-UN members and associate members of the regional commissions. https://www.un.org/en/conf/migration/assets/pdf/Associate-Members-of-the-Regional-Commissions.pdf. Accessed June 16, 2022.

United Nations. (2022). https://data.un.org/_Docs/SYB/PDFs/SYB64_1_202110_Population,percent20Surfacepercent20Areapercent20andpercent20Density.pdf. Accessed June 11, 2022.

United Nations Framework Convention on Climate Change (UNFCCC). (2005). *Climate change: Small island developing states*. UNFCCC Climate Change Secretariat. Accessed June 23, 2022.

US Environmental Protection Agency (EPA). (2022). Overview of Greenhouse Gas Emissions. https://www.epa.gov/ghgemissions/overview-greenhousegases. Accessed June 15, 2023.

US National Oceanic and Atmospheric Administration (NOAA). (2022). NOAA Climate.gov. https://www.climate.gov/. Accessed June 10, 2022.

US NASA. (2022). Deep Concern About Food Insecurity in Eastern Africa. https://earthobservatory.nasa.gov/images/150217/deep-concern-about-foodsecurity-in-eastern-africa. Accessed June 20, 2023.

Vinichenko, V., et al. (2023). Phasing out coal for 2 °C target requires worldwide replication of most ambitious national plans despite security and fairness concerns. *Environmental Research Letters, 18*, 2023. https://doi.org/10.1088/1748-9326/acadfAccessed22June

Williston, B. (2019). *The ethics of climate change*. Routledge.

World Bank, Social Dimensions of Climate Change. (2022). https://www.worldbank.org/en/topic/social-dimensions-of-climate-change#1. Accessed February 17, 2022.

World Bank. (2022). Population total. https://data.worldbank.org/indicator/SP.POP.TOTL. Accessed June 11, 2022.

Worldmiro.com. (2022). The Eora global supply chain database. https://worldmiro.com. Accessed May 31, 2022.

Resources for Keeping Up with Countries' Emissions and Policies

Carbon Tracker. https://carbontracker.org/

European Commission (EC). EDGAR—Emissions Database for Global Atmospheric Research. https://edgar.jrc.ec.europa.eu/

European Union (EU), Copernicus. Climate Change Service. https://climate.copernicus.eu/

Intergovernmental Panel on Climate Change (IPCC). [Periodic reports on many issues]. https://www.ipcc.ch/

International Energy Agency (IEA). [Annual and monthly reports on energy efficiency and other climate-related topics]. https://www.iea.org/

Organisation for Economic Co-operation and Development (OECD). Air and GHG emissions. https://data.oecd.org/air/air-and-ghg-emissions.htm

UK Met Office. [Climate change developments]. https://www.metoffice.gov.uk/about-us/press-office/news/weather-and-climate/

Our World in Data. CO_2 and Greenhouse Gas Emissions. https://ourworldindata.org/co2-and-other-greenhouse-gas-emissions

United Nations Environment Programme (UNEP). World environment situation room. https://wesr-climate.unepgrid.ch/

UNFCCC. GHG data from UNFCCC. https://unfccc.int/process-and-meetings/transparency-and-reporting/greenhouse-gas-data/ghg-data-unfccc/ghg-data-from-unfccc

US National Aeronautics and Space Administration (NASA). Global climate change. https://climate.nasa.gov/

US National Oceanic and Atmospheric Administration (NOAA). https://www.climate.gov/

World Bank. World development indicators: Trends in greenhouse gas emissions. http://wdi.worldbank.org/

World Meteorological Organization (WMO). State of the global climate [annual]. https://public.wmo.int/en/our-mandate/climate/wmo-statement-state-of-global-climate

4 International Agreements

Humanity is conducting an unintended, uncontrolled, globally pervasive experiment whose ultimate consequences could be second only to a global nuclear war.

Toronto World Conference [on the] Changing Atmosphere (1988)

4.1 Introduction

The history of inter-governmental efforts at the international level to understand and address climate change issues extends back more than half a century. There has been periodic progress—initially in the form of scientifically based alarms in the early 1970s—for instance at the First Earth Summit in 1972 and then actions in the form of the 1981 Convention on Long-Range Transboundary Air Pollution, which includes limits on the climate forcer nitrous oxide. The World Conference on the Changing Atmosphere in Toronto in 1988 issued the dire warning above in the chapter's header quotation. Since the early 1990s, the periodic reports of the Intergovernmental Panel on Climate Change (IPCC) have assessed progress in climate science research while the annual Conferences of the Parties (COPS) of the Framework Convention on Climate Change (FCCC) have been conspicuous forums of diplomatic activity.

These and other developments are described in Sects. 4.2 and 4.3. The activities of the many other inter-governmental international organizations that address climate change issues are briefly described in Annex 4.1. Annex 4.2 discusses

T. Brewer, *Climate Change*, https://doi.org/10.1007/978-3-031-42906-4_4

interactions of international trade issues and climate change issues; among them are the treatment of exports and imports in countries' emissions inventories, and the possibilities of including Border Adjustment Measures (BAMs) in Emissions Trading Systems.

4.2 The Beginning (1972–1990)

4.2.1 UN Scientific Conference in Stockholm: The First Earth Summit (1972)

In 1972, an international Scientific Conference in Stockholm that is commonly known as the First Earth Summit adopted a declaration that raised the issue of climate change for the first time in an international inter-governmental conference setting. It warned governments to be mindful of activities that could lead to climate change and [to] evaluate the likelihood and magnitude of climatic effects (Jackson, n.d.). The Conference also called for the convening of a second meeting on the environment, and it established the Governing Council of the United Nations Environment Programme (UNEP), with a secretariat in Nairobi, Kenya.

4.2.2 First World Climate Conference (1979)

The 1979 World Climate Conference in Geneva is often referred to as the First World Climate Conference. Its sponsors included the World Meteorological Organization (WMO), the United Nations Educational, Scientific and Cultural Organization (UNESCO), the Food and Agriculture Organization (FAO), the World Health Organization (WHO), and the United Nations Environment Programme (UNEP). The conference exhorted the world to take full advantage of man's [sic] present knowledge of climate; take steps to improve significantly that knowledge; and foresee and prevent potential man-made changes in climate that might be detrimental to the well-being of humanity. While such exhortations may not seem very precisely focused many decades later, at the time they were at the leading edge of scientific commentary on the subject (Zillman, 2009).

4.2.3 World Climate Research Programme (1980)

Within a year, the WMO led the establishment of the World Climate Research Programme (WCRP), which became a highly interdisciplinary project involving UNESCO, FAO, WHO and UNEP, as well as other institutions focused on climate science research. It still exists and supports research in a wide range of topics (World Climate Research Programme, 2022).

4.2.4 Convention on Long-Range Transboundary Air Pollution (1983) and Gothenburg Protocol (1999)

The Geneva Convention on Long-Range Transboundary Air Pollution (LRTBAP), which entered into force in 1983, was the first international agreement explicitly limiting greenhouse gas emissions (UNECE, 2022). It has expanded beyond its regional focus within the framework of the United Nations Economic Commission for Europe (UNECE). Its original 28 signatories were all in Europe, and its 51 current parties are still mostly in Europe or near it, but they also include Canada and the US.

The 1999 Gothenburg Protocol to the LRTBAP established national emission ceilings for four pollutants (sulfur dioxide, nitrous oxides, volatile organic compounds, and ammonia) and for sources of emissions—including, combustion plants, electricity production, and motor vehicles. An amendment in 2012 made the protocol the first binding agreement to include emission-reduction commitments for fine particulate matter. The amended protocol also specifically includes the short-lived climate pollutant black carbon as very small particulate matter—which is sometimes referred to as "soot" (EU, EUR-LEX, 2022). However, its remit includes the health-threatening air pollutant sulfur dioxide, which happens to be a global coolant; the LRTBAP's net impact on global warming has been a subject of climate science studies (UNECE, 2022).

4.2.5 Toronto Conference (1988)

The World Conference on the Changing Atmosphere in Toronto in 1988 made a strong statement, as reported at the head of this chapter: "Humanity is conducting an unintended, uncontrolled, globally pervasive experiment whose ultimate consequences could be second only to a global nuclear war (UN, 1988)." The Conference also proposed increased resourcing for research and monitoring efforts, support for the work of the Intergovernmental Panel on Climate Change (IPCC) and development of a new international agreement to protect the atmosphere, presumably an agreement that would directly address climate change issues.

4.2.6 Intergovernmental Panel on Climate Change (IPCC) (1988 ...)

The first session of the WMO-UNEP Intergovernmental Panel on Climate Change (IPCC) in Geneva in November 1988 established its three working group structure that still exists: physical science, adaptation and mitigation. In principle the work of the IPCC informs the deliberations at UNFCC meetings, as discussed in Sect. 4.3. Although the IPCC is essentially a scientific organization, its conclusions are subject to review by governmental representatives. Thus, the correspondence between the technicalities of the scientists' conclusions and the summaries of them by the government representatives is not always self-evident.

In 1990, the First IPCC Assessment Report (AR1) underlined the importance of climate change as a challenge with global consequences that required international cooperation (IPCC, 2022a). AR1 played an important role in the creation of the UN Framework Convention on Climate Change (UNFCCC), which has become a key international treaty to reduce global warming and cope with the consequences of climate change. AR2 (1995) provided important material for governments to use prior to adoption of the Kyoto Protocol in 1997. AR3 in 2001 focused attention on the impacts of climate change and the need for adaptation. AR4 in 2007 laid the ground work for a post-Kyoto agreement, and focused on limiting warming to 2 °C. AR5 was finalized between 2013 and 2014, and it provided the scientific input into the Paris Agreement. IPCC AR6 reports (IPCC, 2021, 2022b) were inputs for COP 27.

4.2.7 Second World Climate Conference (1990)

Efforts to raise awareness of the effects of climate change were further advanced at the second World Climate Conference in 1990 (IPCC, 2022a). In its Ministerial Declaration, the Conference stated that climate change was a global problem of unique character for which a global response was required. It called for negotiations to begin on a framework convention without further delay. In fact, negotiations did begin, and they resulted in agreement on the Framework Convention on Climate Change (FCCC).

4.3 The Framework Convention on Climate Change (1992 ...)

In 1992 the UN Conference on Environment and Development in Rio de Janeiro agreed on the Framework Convention on Climate Change (FCCC), which became the central international agreement on climate change issues after it entered into force in 1994 (Bodansky et al. 2017). Among its key provisions, Article 3 provides that the principle of Common but Differentiated Responsibilities and Respective Capabilities (CBDR-RC) applies to all parties to the convention, as follows:

> The Parties should protect the climate system for the benefit of present and future generations of humankind, on the basis of equity and in accordance with their common but differentiated responsibilities and respective capabilities. Accordingly, the developed country Parties should take the lead in combating climate change and the adverse effects thereof.

Issues about the meaning and implementation of the principle became frequent sources of conflict for many years thereafter.

4.3.1 COP 1: Berlin Mandate (1995)

In 1995, the first Conference of the Parties Convention (COP) adopted the Berlin Mandate, which launched talks on a protocol or other legal instrument with commitments that were stronger in developed countries and countries in transition, compared with other countries. The governments of the former two groups of countries committed legally-binding emissions targets according to timetables, while the governments of developing countries did not make any new commitments. The groupings of countries in the UNFCCC process are defined in Box 4.1.

Box 4.1: Groups of Countries in the UNFCCC Process

Annex I Parties include the industrialized countries that were members of the Organisation for Economic Co-operation and Development (OECD) in 1992, plus countries with economies in transition (the EIT Parties), including the Russian Federation, the Baltic States, and several Central and Eastern European States.

Annex II Parties consist of the OECD members of Annex I, but not the EIT Parties. They are required to provide financial resources to enable developing countries to undertake emissions reduction activities under the Convention and to help them adapt to adverse effects of climate change. In addition, they have to "take all practicable steps" to promote the development and transfer of environmentally friendly technologies to EIT Parties and developing countries. Funding provided by Annex II Parties is channeled mostly through the Convention's financial mechanism.

Non-Annex I Parties are mostly developing countries. Certain groups of developing countries are recognized by the Convention as being especially vulnerable to the adverse impacts of climate change, including countries with low-lying coastal areas and those prone to desertification and drought. Others (such as countries that rely heavily on income from fossil fuel production and commerce) feel more vulnerable to the potential economic impacts of climate change response measures. The Convention emphasizes activities that promise to answer the special needs and concerns of these vulnerable countries, such as investment, insurance and technology transfer.

The 49 Parties classified as "least developed countries (LDCs)" by the United Nations are given special consideration under the Convention on account of their limited capacity to respond to climate change and adapt to its adverse effects. Parties are urged to take full account of the special situation of LDCs when considering funding and technology-transfer activities.

Source: UNFCCC (2022a).

4.3.2 Kyoto Protocol (1997)

The Kyoto Protocol was adopted in 1997, but it only entered into force until eight years later in 2005, which was four years after the new US administration in 2001 decided not to support it (Napoli, 2012; Victor, 2004, 2011; Victor & Sabel, 2002). The delays were partly a result of disagreements about how countries could specifically and tangibly fulfill their "common but differentiated responsibilities," as agreed in 1992 at the Rio conference. One way that was already established was through national policies that were to be made public and registered with the UNFCCC secretariat. Each member had to submit an annual emissions inventory report, with an indication of the levels of emissions in relation to its stated targets. The reports were subject to monitoring, reporting and verification (MRV) processes administered by the UNFCCC secretariat in conjunction with the national governments. These MRV processes have evolved since then and are important elements of the administrative support system for the UNFCCC.

Another important innovation included in the Kyoto Protocol was the recognition of three market mechanisms as additional ways for countries to achieve their emission targets (UNFCCC, 2005; Kainou, 2022; Mullins, 2002).

- *International Emissions Trading* was allowed in Article 17 for Annex I countries to sell or buy a variety of tradable financial units in transactions with other Annex I countries.
- The purpose of the *Clean Development Mechanism (CDM)*, according to Article 12, was "to assist Parties not included in Annex I in achieving sustainable development and in contributing to the ultimate objective of the convention, and to assist Parties included in Annex I in achieving compliance with their quantified emission limitation and reduction commitments under Article 3."
- The *Joint implementation (JI)* mechanism in Article 6 allowed transactions between Annex B countries, and such transactions would thus be analogous to CDM transactions between Annex I countries.

In addition to the agreements concerning mitigation targets and methods to achieve them, there was also an agreement to establish an *Adaptation Fund*, which could facilitate international financial assistance to developing countries. Proceeds of the market mechanisms were drawn upon to provide financial support for the fund.

In 2001, COP 7 agreed on the Marrakesh Accords, which provided rules and administrative arrangements to support the Kyoto Protocol (UNFCCC, 2022b). Nevertheless, the Protocol collapsed in 2012, when many governments decided not to extend their expiring commitments Although it had been an ambitious and innovative agreement, many countries did not meet their commitments and thus actual emissions were well above their targets.

4.3.3 COP 21 Paris Agreement (2015)

The Paris Agreement is a legally-binding international treaty that was signed by 196 parties in 2015 and entered into force in 2016 (UNFCCC, 2022b, 2022d; Klein et al. 2017). It is a core, foundational international agreement for much of the international policymaking at subsequent COPS. Among its 29 articles, some are highlighted as folllows:

- ***Long-term temperature goal*** (Art. 2)—The Paris Agreement, in seeking to strengthen the global response to climate change, reaffirms the goal of limiting global temperature increase to well below 2 °C, while pursuing efforts to limit the increase to 1.5°.
- ***Global peaking* and 'climate neutrality'** (Art. 4)—To achieve this temperature goal, Parties aim to reach global peaking of greenhouse gas emissions (GHGs) as soon as possible, recognizing peaking will take longer for developing country Parties, so as to achieve a balance between anthropogenic emissions by sources and removals by sinks of GHGs in the second half of the century.
- ***Mitigation*** (Art. 4)—The Paris Agreement establishes binding commitments by all Parties to prepare, communicate and maintain a nationally determined contribution (NDC) and to pursue domestic measures to achieve them. It also prescribes that Parties shall communicate their NDCs every 5 years and provide information necessary for clarity and transparency. To set a firm foundation for higher ambition, each successive NDC will represent a progression beyond the previous one and reflect the highest possible ambition. Developed countries should continue to take the lead by undertaking absolute economy-wide reduction targets, while developing countries should continue enhancing their mitigation efforts, and are encouraged to move toward economy-wide targets over time in the light of different national circumstances.
- ***Sinks and reservoirs*** (Art. 5)—The Paris Agreement also encourages Parties to conserve and enhance, as appropriate, sinks and reservoirs of GHGs as referred to in Article 4, paragraph 1(d) of the Convention, including forests.
- ***Voluntary cooperation/Market- and non-market-based approaches*** (Art. 6)—The Paris Agreement recognizes the possibility of voluntary cooperation among Parties to allow for higher ambition and sets out principles—including environmental integrity, transparency and robust accounting—for any cooperation that involves internationally transferal of mitigation outcomes. It establishes a mechanism to contribute to the mitigation of GHG emissions and support sustainable development, and defines a framework for non-market approaches to sustainable development.
- ***Adaptation*** (Art. 7)—The Paris Agreement establishes a global goal on adaptation—of enhancing adaptive capacity, strengthening resilience and reducing vulnerability to climate change in the context of the temperature goal of the Agreement. It aims to significantly strengthen national adaptation efforts, including through support and international cooperation. It recognizes that adaptation is a global challenge faced by all. All Parties should engage in

adaptation, including by formulating and implementing National Adaptation Plans, and should submit and periodically update an adaptation communication describing their priorities, needs, plans and actions. The adaptation efforts of developing countries should be recognized.

- ***Loss and damage*** (Art. 8)—The Paris Agreement recognizes the importance of averting, minimizing and addressing loss and damage associated with the adverse effects of climate change, including extreme weather events and slow onset events, and the role of sustainable development in reducing the risk of loss and damage. Parties are to enhance understanding, action and support, including through the Warsaw International Mechanism, on a cooperative and facilitative basis with respect to loss and damage associated with the adverse effects of climate change.
- ***Finance, technology and capacity-building support*** (Art. 9, 10 and 11)—The Paris Agreement reaffirms the obligations of developed countries to support the efforts of developing country Parties to build clean, climate-resilient futures, while for the first time encouraging voluntary contributions by other Parties. Provision of resources should also aim to achieve a balance between adaptation and mitigation. In addition to reporting on finance already provided, developed country Parties commit to submit indicative information on future support every two years, including projected levels of public finance. The agreement also provides that the Financial Mechanism of the Convention, including the Green Climate Fund (GCF), shall serve the Agreement. International cooperation on climate-safe technology development and transfer and building capacity in the developing world are also strengthened: a technology framework is established under the Agreement and capacity-building activities will be strengthened through, inter alia, enhanced support for capacity building actions in developing country Parties and appropriate institutional arrangements. Climate change education, training as well as public awareness, participation and access to information (Art. 12) is also to be enhanced under the Agreement.
- ***Climate change education, training, public awareness, public participation and public access to information*** (Art. 12) is also to be enhanced under the Agreement.
- ***Transparency*** (Art. 13), ***implementation and compliance*** (Art. 15)—The Paris Agreement relies on a robust transparency and accounting system to provide clarity on action and support by Parties, with flexibility for their differing capabilities of Parties. In addition to reporting information on mitigation, adaptation and support, the Agreement requires that the information submitted by each Party undergoes international technical expert review. The Agreement also includes a mechanism that will facilitate implementation and promote compliance in a non-adversarial and non-punitive manner, and will report annually to the CMA.
- ***Global Stocktake*** (Art. 14)—A "global stocktake," to take place in 2023 and every 5 years thereafter, will assess collective progress toward achieving the purpose of the Agreement in a comprehensive and facilitative manner. It will be based on the best available science and its long-term global goal. Its outcome

will inform Parties in updating and enhancing their actions and support and enhancing international cooperation on climate action.

- ***Decision 1/CP.21*** also sets out a number of measures to enhance action prior to 2020, including strengthening the technical examination process, enhancement of provision of urgent finance, technology and support and measures to strengthen high-level engagement. For 2018 a facilitative dialogue is envisaged to take stock of collective progress towards the long-term emission reduction goal of Art. 4. The decision also welcomes the efforts of all non-Party stakeholders to address and respond to climate change, including those of civil society, the private sector, financial institutions, cities and other subnational authorities. These stakeholders are invited to scale up their efforts and showcase them via the Non-State Actor Zone for Climate Action platform. Parties also recognized the need to strengthen the knowledge, technologies, practices and efforts of local communities and indigenous peoples, as well as the important role of providing incentives through tools such as domestic policies and carbon pricing.

The provisions concerning global temperature goals (Article 2) have been highly salient officially and unofficially—and subject to much deliberation. The issue of compensation for "loss and damage" (Article 8) also became active on the COP agendas at the COP 26 and COP 27 meetings in 2021 and 2022.

4.3.4 COP 26 in Glasgow (2021)

The results of COP 26 were a mix of progress on some issues, temporizing on others, and defeat on yet others; in that way, it was typical of many COPs. The most notable developments were agreements about methane and coal.

Global Methane Pledge

The Global Methane Pledge was announced by the EU and US in September 2021, before COP 26 (Global Methane Pledge, 2021). At the beginning of COP 26, there were only a few countries that had agreed to it. By the end of the COP 26 meetings, 104 countries had signed on, and they were responsible for almost half of global methane emissions. However, several major methane emission source countries did not sign on—namely Australia, China, India and Russia.

Methane has accounted for about 30% of global warming since the industrial revolution, with most emissions coming from agriculture and livestock well as oil well flaring in recent years. The goal of the agreement is to reduce methane emissions by 30% by 2030—a reduction that could avoid 0.3 °C of incremental global warming by 2045 (Climate and Clean Air Coalition, 2022). The EU and US announced a package of US$328 million for financial and technical support for implementation of the reduction measures (US Department of State, 2022), including:

- $4 million to support the World Bank Global Gas Flaring Reduction Partnership (GGFR). The United States intends to support the transfer by the World Bank of at least $1.5 million in funding to the GGFR. Germany intends to provide $1.5 million, and Norway intends to provide approximately $1 million to GGFR.
- $5.5 million to support the Global Methane Initiative (GMI). The United States will provide $3.5 million. Guided by the recommendations of the GMI, Canada will contribute $2 million over the next four years, as part of its global climate finance commitment, to support methane mitigation projects in developing countries including in the oil and gas sector.
- Up to $9.5 million from the UNEP International Methane Emissions Observatory to support scientific assessments of methane emissions and mitigation potential in the oil and gas sector that are aligned with the Global Methane Pledge Energy Pathway.
- Up to $40 million annually from the philanthropic Global Methane Hub to support methane mitigation in the fossil energy sector.

Coal

As for an agreement to reduce coal usage, there was last-minute drama when a key phrase that referred to a goal to "phase out" coal-fired plants was changed to "phase down." The change was the result of lobbying by India, China and other coal-dependent countries including the US. (For an analysis of fossil fuel issues in the context of COP 26, see van Asselt & Green, 2022.)

Other Issues

Another agreement, the Glasgow Leaders' Declaration on Forests and Land Use, was signed by more than 100 countries (UN Climate Change, 2022a). The declaration includes a promise to end and reverse deforestation, including in Brazil, by 2030.

Another potentially important development at COP 26 was the announcement of the Just Energy Transition Partnership (JET-P) with South Africa—a program to provide US$ 8.5 billion to South Africa to facilitate the transition away from its current reliance on coal for 90% of its electric power to more climate-friendly sources. The providers of the financial assistance are the EU, France, Germany, UK and US (Carnegie Endowment for International Peace, 2022; Robinson, 2022). The intention of the funders is that it would be a kind of demonstration project that could be applied in other countries, presumably with some adjustments for local political, economic, social and technological conditions. A preliminary discussion of the project had been on the 2021 agenda of the G-7 prior to COP 26 and again on the 2022 agenda of the G-7; that group's meetings were watched for indications that the project was being implemented and that subsequent projects were in the pipeline. However, the G-7 2022 meeting deferred action until COP27 in Sharm El-Sheikh, Egypt.

Otherwise, there were other kinds of progress on familiar topics (UN Climate, 2022b). An adaptation work program was established to define a global goal on adaptation. There was an agreement on mitigation to reduce the gap between projected mitigations and what is needed to achieve the global warming goal of a 1.5 °C increase in the global mean temperature by mid-century. There was a consensus to double (at least) finance for adaptation, and there was a confirmation of the previously pledged US$100 billion dollars of annual contributions from developed to developing countries.

4.3.5 COP 27 in Sharm El-Sheikh (2022)

There were three central issues at COP 27: whether to change the 1.5 °C (2.7 °F) and 2.0 °C (3.6 °F) temperature targets from the Paris Agreement; a carry-over issue from COP 26 about reducing coal use; and loss and damage compensation payments to vulnerable developing countries.

Global Temperature Targets

The issue about global temperature targets was a provision in the Paris Agreement of 2015 at COP 21 that "[reaffirmed] the goal of limiting global temperature increase to well below 2 degrees Celsius, while pursuing efforts to limit the increase to 1.5 degrees" (UNFCCC, 2022b). At COP 27 small island developing countries and other developing countries advocated an agreement to limit global emissions so they would peak by 2025 in order to achieve the 1.5 °C Paris Agreement target. Although there were 80 countries supporting that proposal at COP 27, it was not passed.

Box 4.2 contains excerpts from the IPPC (2023) synthesis of studies of the projections of temperatures in relation to UNFCCC agreements.

Box 4.2: IPCC Synthesis of Studies and IPCC Reports Concerning Temperatures

The UNFCCC, Kyoto Protocol, and the Paris Agreement are supporting rising levels of national ambition. The Paris Agreement, adopted under the UNFCCC, with near universal participation, has led to policy development and target-setting at national and sub-national levels, in particular in relation to mitigation, as well as enhanced transparency of climate action and support (medium confidence). Many regulatory and economic instruments have already been deployed successfully (high confidence). In many countries, policies have enhanced energy efficiency, reduced rates of deforestation and accelerated technology deployment, leading to avoided and in some cases reduced or removed emissions (high confidence). Multiple lines of evidence suggest that mitigation policies have led to several Gt CO_2-eq yr-1 of avoided global emissions (medium confidence). At least 18 countries have sustained absolute production-based GHG and consumption-based CO_2

reductions for longer than 10 years. These reductions have only partly offset global emissions growth (high confidence).

Global GHG emissions in 2030 implied by [UNFCCC] nationally determined contributions (NDCs) announced by October 2021 make it likely that warming will exceed 1.5 °C during the twenty-first century and make it harder to limit warming below 2 °C.

Source: IPCC (2023).

Coal, Oil and Natural Gas

At COP 26 in Glasgow the year before COP 27, there was an agreement on a goal to "phase down" coal-fired power plants; this phrase was a change from "phase out" in the original proposal because of opposition by China, India, the US, and other countries. At COP 27, India proposed that *all* the fossil fuels be phased down—that is oil and natural gas as well as coal. Although more than 80 countries at the conference supported this expansion of fossil fuel phase downs, it was defeated by a group of major oil and gas countries; though Australia and the UK supported the proposal despite being major fossil fuel producers.

Payments to Vulnerable Developing Countries for Loss and Damage

Compensation payments to vulnerable developing countries were already a highly contentious issue for many years before COP 27. However, there was a breakthrough before the beginning of the COP 27 meetings, when the governments of Austria, Belgium, Denmark, Germany, and Scotland (separately from the UK) announced their support for financing such payments with specific amounts of money pledged for specific projects (Yale Environment 360, 2022). At the COP meetings, there was an impasse as the negotiations continued into the hours after the originally scheduled end of the COP meetings. The EU threatened to leave the meetings unless there was a breakthrough in the impasse and made an alternative proposal (Schonhardt & Mathiesen, 2022). An agreement was finally reached after more than 36 hours of overtime negotiations in the form of two "decisions" embedded in the *Funding arrangements for responding to loss and damage associated with the adverse effects of climate change, including a focus on addressing loss and damage* (see Box 4.3).

Box 4.3: Loss and Damage Decisions at COP 27

Decide to establish new funding arrangements for assisting developing countries that are particularly vulnerable to the adverse effects of climate change, in responding to loss and damage, including with a focus on addressing loss and damage by providing and assisting in mobilizing new and additional resources, and that these new arrangements complement and include sources, funds, processes and initiatives under and outside the Convention and the Paris Agreement;

Also decide, in the context of establishing the new funding arrangements referred to in paragraph 2 above, to establish a fund for responding to loss and damage whose mandate includes a focus on addressing loss and damage. …

Source: UNFCCC (2022c).

Thus, a new fund specifically focused on helping developing countries that suffer damage and loss from climate change with "new and additional resources" was to be established. However, there was no indication of the amounts of money to be contributed or by whom; nor was there an indication of the specific institutional arrangements that would be made to manage the fund. These and many other issues were deferred to the work of a transitional committee, which would make recommendations to COP 28 the following year.

Sources for COP 27: Borenstein et al. (2022), Brahic (2022), Euractif (2022), Harvey (2022), Hodgson (2022), International Institute for Sustainable Development (2022), McGrath (2022), Michaelowa (2022), Reuters (2022), Schonhardt and Mathiesen (2022), Stavins (2022), UNFCCC (2022e), World Resources Institute (2022).

4.4 Conclusion

There has been much expansion of international organizational programs concerning climate change since the 1990s, and the programs will reduce the impacts of climate change to less than they would have been otherwise; however, as of the early 2020s, they are nevertheless inadequate to meet the challenges posed by increasingly ominous climate change science reports as well as widespread and catastrophic extreme weather events. Thus, it has been suggested that:

> Given [the] risks, it is shocking that the multilateral system has failed to respond more forcefully and has instead merely tinkered at the margins. … It is time to govern the world as if the earth mattered. What the world needs is a paradigm shift in … international relations - a shift that is rooted in ecological realism and that moves cooperation on shared environmental threats to center stage. Call this new worldview "planetary politics" (Patrick, 2021).

Annex 4.1: Climate Change Programs in International Organizations

The following brief descriptions include organizational mission statements and other self-descriptive materials.

Climate and Clean Air Coalition (CCAC)—https://www.ccacoalition.org/
The Climate and Clean Air Coalition is a voluntary partnership of governments, intergovernmental organizations, businesses, scientific institutions and civil society organizations committed to improving air quality and protecting the climate through actions to reduce short-lived climate pollutants.
Food and Agriculture Organization (FAO)—https://www.fao.org/
Office of Climate Change, Biodiversity and Environment (OCB).
International Energy Agency (IEA)—https://www.iea.org/
Areas of work include: Promoting energy efficiency; Ensuring energy security; Technology collaboration; Data and statistics.
International Labour Organization (ILO)—https://www.ilo.org/
Many publications about climate change and employment.
Organization for Economic Cooperation and Development (OECD)—https://www.oecd.org/climate-change/
The OECD supports and helps drive higher levels of ambition and tangible outcomes—on mitigation, adaptation and resilience, and financing—that better align with the collective goals of the Paris Agreement. This includes enhanced support for climate action in developing countries, including in the least developed and most vulnerable countries which are heavily impacted by climate change, yet lack access to tools required to support a transition to net-zero emissions.
United Nations Conference on Trade and Development (UNCTAD)—https://unctad.org/topic/trade-and-environment/climate-change
We support efforts to promote sustainability, adapt and build resilience against the climate emergency.
United Nations Environment Programme (UNEP)—https://www.unep.org/explore-topics/climate-action
World Bank—https://www.worldbank.org/en/topic/climatechange
The biggest multilateral funder of climate investments in developing countries.
Regional Development Banks
African Development Bank—https://www.afdb.org/en/topics-and-sectors/sectors/climate-change
Asian Development Bank—https://www.adb.org/what-we-do/themes/environment/main
European Bank for Reconstruction and Development—https://www.ebrd.com/sustainable-resources-and-climate-change.html
Inter-American Development Bank—https://www.iadb.org/en/climate-change/creating-innovative-development-opportunities

United Nations Framework Convention on Climate Change (UNFCCC)—https://unfccc.int/
Base agreement for the annual Conferences of the Parties (COPs). Secretariat in Bonn. Website for archives of documents, decisions and other resources.
World Health Organisation (WHO)—https://www.who.int/health-topics/climate-change#tab=tab_1
Co-sponsors with World Meteorological Organization (WMO) a global knowledge platform dedicated to climate and health (climahealth.info).
World Meterological Organisation (WMO)—https://public.wmo.int/en/our-mandate/climate
Publishes annual *State of the Global Climate.* Programs and projects include: Global Framework for Climate Services; Climate Resilience and Adaptation. Co-sponsors with World Meteorological Organization (WMO) a global knowledge platform dedicated to climate and health (climahealth.info).

Annex 4.2: International Trade Issues

Two sets of trade related issues are especially relevant to this chapter's focus on international agreements. One is a mix of technical accounting issues and policy issues concerning countries' annual emissions inventories and reports to the FCCC. The other is a mix of trade economics and trade laws related to the establishment of Emissions Trading Systems (ETSs). There is a third category of "others" – that is, a wide variety of issues that have been on the WTO trade agenda, but not so much on the climate change agenda. These three categories are considered in turn.

Exclusion of Exports and Imports in Countries' FCCC Emissions Inventory Allocations
A central, continuing and contentious methodological issue in international climate change agreements is whether to attribute emissions to *exporting countries' production of the exports* or *importing countries' consumption of them.* It is a specific policy issue in the reporting of countries' inventories of their emissions in order to gauge progress—or its opposite—toward meeting Paris Agreement temperature targets and otherwise mitigating emissions. Nationally determined contributions (NDCs) are at the heart of the Paris Agreement and the achievement of long-term goals. NDCs embody efforts by each country to reduce national emissions and adapt to the impacts of climate change. The Paris Agreement (Article 4, paragraph 2) requires each Party to prepare, communicate and maintain successive nationally determined contributions (NDCs) that it intends to achieve, and it is expected to pursue domestic mitigation measures in order to achieve the NDCs. However, the reports include exporting countries' emissions from producing, while importing countries do not report their consumption levels. This practice violates a widely accepted principle that consumers should be responsible for the emissions.

Border Adjustment Measures (BAMs) for Emission Trading Systems (ETSs)
Among policy and legal specialists, there has been intensive interest in the intersection of WTO international trade issues with possible border adjustment measures (BAMs) that could be instituted as elements of emission trading systems (ETSs). Interest in BAM issues has been especially strong in Europe in regard to the all-ready established EU ETS. Two different but related issues concern (a) the economics of the competitive position of industries in the countries in an ETS and (b) the legality of BAMs in international trade law and thus exposure to the risk of a WTO dispute settlement decision that could restrict the use of BAMs. Because of the significance and technicalities of the economic and legal issues, there are many studies of BAM issues (for instance, Brewer, 2004; Mehling 2022). Because these are on-going issues as of this writing, readers are encouraged to consult EC and WTO sources, as well as Reuters, AP News and Politico.eu.

Other Trade-Climate Issues
There is a long list of issues about other kinds of climate-trade interactions, and these have been the subject of WTO staff papers (2011, 2021) that focus on the details of WTO rules, trade measures adopted by its members, and a combination of legal, economic and climate change considerations that are inherent in balanced and thorough analysis of the issues. Issue areas of special significance at the WTO include subsidies, technical standards, government procurement, and local content requirements for foreign direct investments; it should also be noted that many issues apply to trade in services as well as goods. The climate-trade agenda is thus broad and consequential for both climate and trade.

The WTO secretariat noted (WTO, 2022a) in an analysis of trade-climate issues that "In the Marrakesh Agreement establishing the WTO, members established a clear link between sustainable development and disciplined trade liberalization—in order to ensure that market opening goes hand in hand with environmental and social objectives." It highlighted several specific areas of special importance:

- The Committee on Trade and Environment
- The Committee on Technical Barriers to Trade
- Liberalizing environmental goods
- Liberalizing environmental services
- Negotiations on agriculture
- Market access for non-agricultural goods.

There have been many occasions when climate issues have been on the agenda of WTO committees, councils and other forums (WTO, 2022b):

- 16–17 September 2021, Trade and Environmental Sustainability Structured Discussions (TESSD) meeting—Members advance work on trade and environmental sustainability ministerial declaration

- 30 March 2021, Meeting of the Committee on Trade and Environment (CTE)—WTO members discuss measures to tackle climate change and strengthen sustainability
- 23 June 2021, Meeting of the Committee on Trade and Environment (CTE)—Members discuss sustainable food, fisheries and new environmental initiatives
- 12 and 16 November 2020, Committee on Market Access—Brexit, EU's carbon border adjustment mechanism on the agenda of the Market Access Committee
- 10–11 June 2020, Council for Trade in Goods—Goods Council considers EU plans for carbon taxes on certain imports
- 30 September 2019, Council for Trade in Services Special Session—WTO members engage in exploratory talks on market access for environmental services.

Subsidies of low-carbon alternatives to fossil fuels have become a trade issue. For instance, a trade dispute concerning climate change policies developed in 2022 when …

> The European Union … initiated a [WTO] dispute with the United Kingdom due to subsidies London is offering to promote low carbon energy generation projects… [T]he EU alleges that the United Kingdom is acting inconsistently with international trade rules by making local content a criterion of eligibility for, and payment of, subsidies for low carbon energy generation projects (Reuters, 2022).

Another kind of trade policy that has become especially conflictual is the use of local content requirements as a barrier to imports in conjunction with subsidies of domestic production of climate-friendly products such as electric vehicles. There are several such provisions in the major 2022 US climate change legislation in the form of the "Inflation Reduction Act," which includes hundreds of billions of dollars in climate-friendly subsidies.

Questions to Ponder

1. What do you think have been the most important successes and failures of international agreements concerning climate change? Why?
2. If you were the Secretary General of the United Nations, what would be the themes of your speeches about climate change?
3. If you were the ambassador of your own home country to the United Nations General Assembly, what would you say about climate change that reflects the official views of your home country's government? (These would not necessarily be your own personal views, of course, since you would be speaking on behalf of the government. Use your imagination as well as references to government documents.)
4. If you could reform the entire system of multilateral organizations to be more effective in addressing climate change issues, what would you do?

References

Bodansky, D., Brunnée, J., & Rajamani, L. (2017). *International climate change law*. Oxford: Oxford University Press.

Borenstein, S., Magdy, S., & Jordans, F. (2022). UN climate deal. https://apnews.com/.com/hub/climate-and-environment. Accessed November 20, 2022.

Brahic, B. (2022). What happened at COP27? *The Economist*. https://www.economist.com/films/2022/11/20/what-happened-at-cop27. Accessed November 24, 2022.

Brewer, T. (2004). The WTO and the Kyoto protocol: Interaction issues. *Climate Policy, 4*(1), 3–12. https://doi.org/10.1080/14693062.2004.9685506

Carnegie Endowment for International Peace. (2022). South Africa's just energy transition partnership. https://carnegieendowment.org/2022/09/28/south-africa-s-just-energy-transition-partnership-model-for-climate-financing-for-africa-event-7948. Accessed November 25, 2022.

Climate and Clean Air Coalition (CCAC). (2022). Global methane assessment. https://www.ccacoalition.org/en/resources/global-methane-assessment-full-report. Accessed November 25, 2022.

Euractif. (2022). Berlin, Paris fume after 'disappointing' COP results. Accessed at https://www.euractiv.com/section/politics/news/berlin-paris-fume-after-disappointing-cop-results/?utm_source=piano&utm_medium=email&utm_campaign=26669&pnespid=7r07CSlcMrEcyKnRuDOvTZaWok32UpZ4M_Wineo29xRmHSZON62x7mnJLmTj7ScB69IS0sxe on November 21, 2022.

EU, EUR-LEX. (2022). Geneva convention on long-range transboundary air pollution. https://eur-lex.europa.eu/legal-content/EN/TXT/?uri=LEGISSUM%3Al28162#:~:text=Long%2Drange%20transboundary%20air%20pollution%20is%20defined%20as%20the%20release,groups%20of%20sources%20cannot%20be. Accessed March 23, 2022.

Global Methane Pledge. (2021). Global methane pledge. Accessed at https://www.globalmethanepledge.org/ on November 25, 2022.

Harvey, F. (2022). World still 'on brink of climate catastrophe' after Cop27 deal. https://www.theguardian.com/environment/2022/nov/20/world-still-on-brink-of-climate-catastrophe-after-cop27-deal?utm_term=637b765432dca98693664c05d6291a58&utm_campaign=GuardianTodayUS&utm_source=esp&utm_medium=Email&CMP=GTUS_email. Accessed November 21, 2022.

Hodgson, C. (2022). COP27 ends in tears and frustration. https://www.ft.com/content/03d7609b-decc-40fc-8029-a6cdebb11752. Accessed November 20, 2022.

International Institute for Sustainable Development (IISD). (2022). *Earth Negotiations Bulletin*. Summary of the Sharm El-Sheikh climate change conference: 6–20 November 2022. Accessed at https://enb.iisd.org/sharm-el-sheikh-climate-change-conference-cop27 on November 23, 2022.

Intergovernmental Panel on Climate Change (IPCC). (2021) *Climate change 2021: The physical science basis*. Working Group I contribution to the IPCC sixth assessment report. Accessed at https://www.ipcc.ch/report/sixth-assessment-report-working-group-i/ on November 24, 2022.

Intergovernmental Panel on Climate Change (IPCC). (2022a). History of the IPCC. https://www.ipcc.ch/about/history/ Accessed November 24, 2022.

Intergovernmental Panel on Climate Change (IPCC). (2022b). Climate change 2022: Impacts, adaptation and vulnerability. https://www.ipcc.ch/report/sixth-assessment-report-working-group-ii/. Accessed November 24, 2022.

Intergovernmental Panel on Climate Change (IPCC). (2023). Synthesis report of the IPCC sixth assessment report (AR6), summary for policymakers. https://report.ipcc.ch/ar6syr/pdf/IPCC_AR6_SYR_SPM.pdf. Accessed March 21, 2023.

Jackson, P. (n.d.). From Stockholm to Kyoto: A brief history of climate change. *UN Chronicle*. Accessed at https://www.un.org/en/chronicle/article/stockholm-kyoto-brief-history-climate-change on November 20, 2022.

Kainou, K. (2022). *Collapse of the clean development mechanism scheme under the Kyoto protocol and its spillover: Consequences of 'carbon panic'*. Centre for Economic Policy Research (CEPR). Accessed at https://cepr.org/voxeu/columns/collapse-clean-development-mechanism-scheme-under-kyoto-protocol-and-its-spillover on October 24, 2022.

Klein, D., et al. (2017). *The Paris agreement on climate change*. Oxford University Press.

McGrath, M. (2022). Climate change: Five key takeaways from COP27. Accessed at https://www.bbc.com/news/science-environment-63693738. Accessed November 20, 2022.

Mehling, M., van Asselt, H., Droege, S., & Das, K. (2022). *The form and substance of international cooperation on border carbon adjustments*. Cambridge: Cambridge University Press.

Michaelowa, A. (2022). Digesting COP27. www.perspectives.cc. Accessed November 20, 2022.

Mullins, F. (2002). Joint implementation institutions: Implementing JI at the national level. OECD. https://www.oecd.org/czech/2766355.pdf. Accessed October 24, 2022.

Napoli, C. (2012). Understanding Kyoto's failure. *SAIS Review of International Affairs, 32*, 2, 183–196. Accessed at https://muse-jhu-edu.proxy.library.georgetown.edu/article/493430/pdf on October 24, 2022.

Patrick, S. M. (2021). The international order isn't ready for the climate crisis: The case for a new planetary politics. *Foreign Affairs*, November/December 2021, 166–176.

Reuters. (2022). COP27 deal delivers landmark on 'loss and damage', but little else. https://www.reuters.com/business/cop/countries-agree-loss-damage-fund-final-cop27-deal-elusive-2022-11-20/. Accessed November 20, 2022.

Robinson, M. (2022). The G7's chance for a just climate transition. *Politico*, 26 June. https://www.politico.eu/article/g7-chance-just-climate-transition/. Accessed June 27, 2022.

Schonhardt, S., & Mathiesen, K. (2022). How vulnerable countries finally got a fund for climate damage, Politico. Accessed at https://www.politico.eu/article/how-vulnerable-countries-finally-got-a-fund-for-climate-damage-cop27/ on September 24, 2022.

Stavins, R. (2022). What really happened at COP27 in Sharm El-Sheikh? An economic view of the environment. Accessed at https://www.robertstavinsblog.org/2022/11/22/what-really-happened-at-cop27-in-sharm-el-sheikh/ on November 25, 2022.

UN. (1988). Proceedings, world conference, Toronto, Canada June 27–30, 1988: The changing atmosphere: Implications for global security. https://digitallibrary.un.org/record/106359. Accessed February 16, 2022.

United Nations, Climate Change (2022a) COP26: Pivotal progress made on sustainable forest management and conservation. Accessed at https://unfccc.int/news/cop26-pivotal-progress-made-on-sustainable-forest-management-and-conservation on November 25, 2022.

United Nations, Climate Change. (2022b). COP 26: Together for our planet. Accessed at https://www.un.org/en/climatechange/cop26 on November 25, 2022.

UN. (2021). COP26 reaches consensus on key actions to address climate change. UN climate press release. Accessed at https://unfccc.int/news/cop26-reaches-consensus-on-key-actions-to-address-climate-change on June 27, 2022.

United Nations Economic Commission for Europe (UNECE). (2022). The convention and its achievements. Accessed at https://unece.org/convention-and-its-achievements on November 25, 2022.

United Nations Framework Convention on Climate Change (UNFCCC). (2005). Mechanisms under the Kyoto Protocol. Accessed at https://unfccc.int/process/the-kyoto-protocol/mechanisms on October 23, 2022.

United Nations Framework Convention on Climate Change (UNFCCC). (2022a). Documents. Accessed at https://unfccc.int/documents on February 16, 2022.

United Nations Framework Convention on Climate Change (UNFCCC). (2022b). Climate change. The guidelines to implement the Kyoto protocol: The Marrakesh accords and the 5, 7 &

8 implications. https://unfccc.int/process/transparency-and-reporting/reporting-and-review-under-the-kyoto-protocol/overview/background-and-resources/the-guidelines-to-implement-the-kyoto-protocol-the-marrakesh-accords-and-the-578-implications. Accessed October 25, 2022d.

United Nations Framework Convention on Climate Change (UNFCCC). (2022c). Funding arrangements for responding to loss and damage associated with the adverse effects of climate change, including a focus on addressing loss and damage. https://unfccc.int. Accessed November 21, 2022.

United Nations Framework Convention on Climate Change (UNFCCC). (2022d). Climate change UNFCCC process-and-meetings, The Paris Agreement. https://unfccc.int/process-and-meetings#:a0659cbd-3b30-4c05-a4f9-268f16e5dd6b. Accessed March 27, 2022.

United Nations Framework Convention on Climate Change (UNFCCC). (2022e). COP 27—UN climate change conference. https://unfccc.int/. Accessed November 24, 2022.

US Department of States. (2022). U.S-EU joint press release on the global methane pledge energy pathway. https://www.state.gov/u-s-eu-joint-press-release-on-the-global-methane-pledge-energy-pathway/. Accessed November 25, 2022.

van Asselt, H., & Green, F. (2022). COP26 and the dynamics of anti-fossil fuel norms. *WIREs Climate Change*. https://wires.onlinelibrary.wiley.com/doi/pdfdirect/10.1002/wcc.816. Accessed November 22, 2022.

Victor, D. (2004). *The collapse of the Kyoto protocol and the struggle to slow global warming*. Princeton University Press.

Victor, D. (2011). *Global warming gridlock: Creating more effective strategies for protecting the planet*. Cambridge University Press.

Victor, D., & Sabel, C. (2002). *Fixing the climate: Strategies for an uncertain world*. Princeton University Press.

World Climate Research Programme. (2022). About WCRP. https://www.wcrp-climate.org/about-wcrp/wcrp-overview. Accessed May 12, 2022.

World Resources Institute. (2022). COP27 resources hub. https://www.wri.org/un-climate-change-conference-resource-hub. Accessed November 20, 2022.

World Trade Organization (WTO). (2022a). Trade and climate change. WTO meetings at which climate change has been raised. https://www.wto.org/english/tratop_e/envir_e/climate_intro_e.htm. Accessed November 20, 2022a.

World Trade Organization (WTO). (2022b). The activities of the WTO and the challenge of climate change. https://www.wto.org/english/tratop_e/envir_e/climate_challenge_e.htm. Accessed March 31, 2022.

Yale Environment 360. (2022). Austria joins Germany, Denmark, and other rich nations pledging millions for climate damages. https://e360.yale.edu/digest/austria-loss-and-damage-climate-change. Accessed November 25, 2022.

Zillman, J. (2009). A history of climate activities. Accessed at https://public.wmo.int/en/bulletin/history-climate-activities on November 25, 2022.

Resources for Keeping Track of Developments in International Agreements

Climate analytics. Climate action tracker. https://climateanalytics.org/projects/climate-action-tracker/

International Institute for Sustainable Development. (IISD). https://www.iisd.org/focus-areas/climate

International Organization. (Journal). https://www.cambridge.org/core/journals/international-organization
United Nations Framework Convention on Climate Change. (UNFCCC). https://unfccc.int/

Part III
Sectors

Fossil Fuels 5

A year ago [1996], the Second Report of the Inter-Governmental Panel on Climate Change was published. That report and the discussion which has continued since its publication, shows that there is mounting concern about two stark facts. The concentration of carbon dioxide in the atmosphere is rising, and the temperature of the earth's surface is increasing.

John Browne, British Petroleum Chief Executive

Stanford University Speech, 19 May 1997 (Browne, 1997)

5.1 Introduction

This chapter focusses on the production and distribution of oil, natural gas, and coal. The chapter addresses the central climate change issues—namely, the sources, volumes, and effects of fossil fuel emissions; and how to reduce the emissions and their effects. There are many technological innovations offering opportunities to change production and distribution systems. More broadly, the socio-economic and political issues of phasing down the industry are especially contentious. These issues arise from the widespread agreement among climate change experts that significant reductions of fossil fuel production and consumption must be made this decade in order to avoid the most catastrophic consequences of climate change (IEA, 2023a, 2023b, 2023c; IPCC, 2022a, 2022b, 2023).

T. Brewer, *Climate Change*, https://doi.org/10.1007/978-3-031-42906-4_5

Section 5.2 analyzes the production and consumption patterns and trends of fossil fuels. Fossil fuel emissions, with an emphasis on methane, are the focus of Sect. 5.3. Section 5.4 addresses issues about public positions taken by corporations in the industry. Section 5.5 discusses government subsidies of fossil fuels. Section 5.6 describes the wide range of mitigation policies that can be applied to fossil fuels. Section 5.7 summarizes the chapter.

Fossil fuels and alternative fuels for the electric power sector and the transportation sector are considered in Chaps. 6 and 7. Those sectors are of course major users of fossil fuels and are themselves significant sources of emissions from burning fossil fuels.

5.2 Fossil Fuel Production

The data in Table 5.1 for 2010 are straight-forward and require no elaboration. However, it should be noted that 2020 was in the midst of the covid pandemic, during which year-to-year fossil fuel production volumes fluctuated. Also note that the data for 2030 are projections of "stated policies"—that is, IEA projections based on IEA scenarios computed from models that included the "stated policies" of national governments.

Section (a) in Table 5.1, with data for oil volumes, indicates a world increase of 5.8% from 2010 to 2020, and projects a 12.1% increase from 2020 to 2030. The biggest rate of increase is for North America (i.e. Canada and the US), where a doubling of 2010 volumes is projected for 2030. The analogous projection for the Middle East is an increase of one-third by 2030 compared with 2010.

Section (b) in Table 5.1, with natural gas volumes, projects a world increase of 9.4% from 2020 to 2030, following a 22.1% increase from 2020 to 2030. As with oil, the largest increases are for North America, with 2030 projected to be 58.2% higher than 2010.

The trends in Section (c) in Table 5.1 for coal volumes are dramatically different at the world and regional levels. At the world level, after an increase of 4.3% from 2010 to 2020, the total projected for 2030 is 5.5% less than the actual 2020 volume. However, the Asia Pacific region—which includes Australia, China, India and other countries, and which is the dominant region in coal production—increased by one-fourth from 2010 to 2020 but is projected to remain at virtually the same level from 2020 to 2030.

In sum, whereas world volumes of oil and natural gas have been increasing and are expected to continue to increase, coal volumes may have peaked by the early 2020s.

Table 5.1 Global and regional production volumes

Producers	Million Barrels per Day (mb/d)		
	2010	2020	2030 projections of "stated policies"
(a) Oil			
World	83.4	89.1	99.9
OPEC[a]	33.3	30.8	35.9
Middle East	25.4	27.6	33.9
North America	14.2	23.9	28.6
Eurasia[b]	13.4	13.4	11.9
Africa	10.2	7.0	7.0
Asia Pacific	8.4	7.5	5.3
Central and South America	7.4	5.9	9.0
Europe [c]	4.4	3.8	3.6
Producers	**Billion cubic meters equivalent (bcm)**		
	2010	2020	2030 projections of "stated policies"
(b) Natural gas			
World	3274	3996	4372
North America	811	1164	1283
Eurasia[c]	807	911	831
Asia Pacific	488	639	694
Middle East	463	646	853
Europe[b]	341	245	247
Africa	203	237	313
Central and South America	160	154	149
Producers	**Million tonnes of coal equivalent (Mtce)**		
	2010	2020	2030 projections of "stated policies"
(c) Coal			
World	5235	5459	5149
Asia Pacific[d]	3487	4299	4282
North America	818	409	188
Europe[b]	331	185	126
Eurasia[c]	309	400	323
Africa	210	211	188

(continued)

Table 5.1 (continued)

Producers	Million tonnes of coal equivalent (Mtce)		
	2010	2020	2030 projections of "stated policies"
Central and South America	79	53	41
Middle East	1	1	1

Source Excerpted and compiled by Thomas L. Brewer from International Energy Agency (IEA) (2023c: Table A.7, Table A.11 and Table A.13)
[a]As of 1 April 2023, there were 13 members of OPEC: Algeria, Angola, Congo, Equatorial Guinea, Gabon, Iran, Iraq, Kuwait, Libya, Nigeria, Saudi Arabia, United Arab Emirates, and Venezuela (Ecuador, Indonesia, and Qatar participated for varying periods in the past)
[b]Includes Norway, UK and other non-EU members plus all 27 EU members
[c]Includes Russia
[d]Includes Australia, China, India and other countries

5.3 Emissions

When analyzing fossil fuel emissions, it is important to distinguish among the three "scopes," which are defined by the Greenhouse Gas Protocol (2023) as:

- Scope 1, directly from a fossil fuel production, distribution or refinery facility;
- Scope 2, from the use of the fuel to produce electricity, steam, heat or cooling;
- Scope 3, other indirect uses.

5.3.1 Trends and Components in Fossil Fuel Emissions

There were three distinct periods in fossil fuel emissions from 1750 before the industrial revolution until the early 2020s: 1750–1850, when there were almost no emissions; 1850 until 1950, when coal emissions rose steadily while oil and gas emission began to emerge; and 1950 to the early 2020s, when coal, oil and gas all increased dramatically (Ritchie et al. 2020). By 2022, world CO_2 emissions from fossil fuels included 12 billion tonnes from oil, 8 billion tonnes from natural gas, and 15 billion tonnes from coal (Global Carbon Project, 2023; Ritchie & Roser, 2020).

5.3.2 Methane Emissions

Among the several types of fossil fuel emissions, those of methane are especially significant because they have occurred at increasingly high levels for decades and because they are highly potent and short-lived climate forcing agents. Their 20-year Global Warming Potential (GWP) is 82.5 times greater per tonne than CO_2, and their 100-year GWP is 29.8 times greater than CO_2. Reducing human-caused methane emissions is one of the most cost-effective strategies to rapidly reduce the rate of warming.

Especially dramatic cases of methane "super-emitting events" in Turkmenistan have made it a world leader in methane emissions (see Box 5.1).

Box 5.1: Super-Emitting Events in Turkmenistan

There are two main sources of methane emissions in Turkmenistan in central Asia: its western fossil fuel fields and its eastern fossil fuel fields. Turkmenistan exports most of its natural gas production to China and is second only to Australia, as a foreign supplier for China.

Turkmenistan's methane super-emitting events has made it number one in the world. As with other countries, a switch from flaring methane emissions to venting them has been one reason for the uptick from 2019. The number of such events approximately doubled from 2021 to 2022. In 2022, the western field emitted 2.6 thousand tonnes of methane and the eastern field emitted 1.8 thousand tonnes. *The total 4.4 thousand tonnes of methane emissions of the two fields in CO_2-equivalents using a 20-year GWP was 366 thousand tonnes—which was more than the annual national CO_2-equivalents emissions of the UK.*

The prospective development of off-shore fields is likely to increase the levels of emissions. As of early-2023, the Turkmenistan government had not joined either the Global Methane Pledge or the Joint Declaration from Energy Importers and Exporters on Reducing Greenhouse Gas Emissions from Fossil Fuels. (See Annex 5.1 for the content of the two agreements.) However, in June 2023 the president announced that Turkmenistan would sign the Global Methane Pledge, which includes a pledge to reduce methane emissions by 30% by 2030. He also indicated the his government would join foreign partners in projects and collaborate with the International Methane Emissions Observatory (IMEO).

Sources: Compiled by the author from Carrington (2023), IEA (2023a, 2023b, 2023c), and Irakulis-Loitxate (2022).

Methane monitoring methods and issues

Widely used methane data are from the International Energy Agency (IEA), which reports a professionally prepared data set. However, it should be noted that the IEA has itself alerted users that its data are typically underestimates obtained from the fossil fuel industry firms (Box 5.2). (Agriculture is also a significant source of methane emissions, as are waste sites.)

Incomplete information about actual emission levels and a lack of awareness of the cost-effectiveness of abating emissions is a key barrier to reducing methane emissions. In response, a growing number of recent initiatives aim to measure and report current and historical emissions from facilities, production types and countries. Yet these assessments remain incomplete; most countries and regions still have little or no measurement-based data and the data they provide often requires careful processing. These gaps highlight the need for robust and transparent data evaluation and harmonization of estimates.

Box 5.2: Opportunities for Reducing Methane Emissions
[Curbing methane] emissions is the most effective means available for limiting global warming in the near term. Emissions from fossil fuel operations present a major opportunity in this respect, since the pathways to reduction are both clear and cost-effective. Fossil fuel operations generated close to 120 Mt of methane in 2020—nearly one-third of all methane emissions from human activity. The scope for reducing these emissions is enormous. This is particularly true in the oil and gas sector, where it is possible to avoid more than 70% of current emissions with existing technology, and where around 45% could be avoided at no net cost.

Source: IEA (2022d); also see IEA (2023b) and UNEP/CCAC (2021a, 2021b).

5.4 Issues About Industry Public Positions

There have been issues about the fossil fuel industry´s private and public positions concerning the scientific evidence about climate change. However, in contrast to some other firms, the excerpt in the chapter heading from a speech in 1997 by John Brown, CEO of British Petroleum at the time, represented an early public recognition of the existence of climate change from one of the world's major oil companies (Browne, 1997).

Ten years later a senior scientist at Exxon—also among the biggest oil companies—wrote an internal memo in July 2007 to the corporation's management committee saying that "there is general scientific agreement that the most likely manner in which mankind is influencing the global climate is through carbon dioxide release from the burning of fossil fuels …." (Hall, 2015; also see Mann, 2012; Orestes & Conway, 2010; Supran & Oreskes, 2017, 2020; Supran et al. 2023).

The results of climate science research were brought directly to the attention of ExxonMobil oil corporation executives by the British Royal Society in a 2006 letter (The Royal Society, 2006a, 2006b). ExxonMobil's response is in The Royal

Society (2006c). The essence of the exchange was that The Royal Society's letter noted to ExxonMobil that the "… statements in your documents were not consistent with the scientific literature that has been published on this issue." The ExxonMobil response was that The Royal Society had "… incorrectly and unfairly described our company and our approach to climate change." (See chapter Annex 5.2 for more about this exchange.)

More recently, some oil corporations' public statements about their actions to mitigate the risks of climate change have been described as "greenwashing"—i.e. making claims about their responses to climate change that are not accurate (Brower, 2023). As a consequence of these developments, issues about some oil corporations' advertisements and other public statements have been introduced in court systems in the UK, US, and other countries (Ricketts, 2023).

5.5 Fossil Fuel Subsidies

Estimates of the annual global costs of governments' fossil fuels subsidies vary widely from hundreds of billions of US dollars to nearly 6 trillion. One of the reasons is that there are many kinds of subsidies—including for fossil fuel producers; electricity producers that use fossil fuels; providers and consumers of transportation services; steel, cement, and chemical plants; heating in the buildings sector; and fossil fuels in the agriculture sector. Some estimates focus directly on the fossil fuel industry itself; others include direct and indirect subsidies in one or more of the many categories of users in other sectors.

Another contributing factor to the differences is that estimates vary for the "price gap" that is commonly used as the key metric. That number is the gap between free-market reference prices and the prices changed to consumers. Yet another factor is foreign exchange rates used to convert the global metric into US dollar equivalents—sometimes, but not always, a "real exchange rate" that reflects differences and changes in inflation rates and therefore purchasing power. Related to that, the US dollar amounts may be reported in nominal terms or inflation-adjusted terms over time.

Nevertheless, there is abundant evidence about two features of the subsidies. One is that they have generally been increasing over time—in "real" inflation-adjusted terms, not only "nominal" terms. The second feature is that the subsidy levels change over time. Specifically, as the prices of fossil fuels increase, the costs of subsidies increase because governments respond to the economic and political pressures when consumer prices increase. Contrariwise, the costs of subsidies decline when fossil fuel prices decline and governments can reduce these program budgets.

With these caveats in mind, it is feasible to review some specific data. A useful source is the Fossil Fuel Subsidy Tracker (2023), which draws upon data from the IEA, IMF, and OECD. For 2021, it reported a total of USD 731.65 billion for "82 major economies." Almost half was for petroleum, roughly one-fourth for each of natural gas and "end use electricity," and about 3% for coal.

Based on a more encompassing notion that included "implicit" subsidies as well as "explicit" subsidies, an IMF (2023) estimate for a global total in 2020 was USD 5.9 trillion. Explicit subsidies were less than USD 1 trillion, as they were for the years before and after 2020. The distinction between the two types, which is clearly important in any analysis of the subsidies, is defined in Box 5.3.

Box 5.3: Explicit and Implicit Subsidies

Explicit subsidies occur when the retail price is below a fuel's supply cost. For a non-tradable product (e.g., coal), the supply cost is the domestic production cost, inclusive of any costs to deliver the energy to the consumer, such as distribution costs and margins. In contrast, for an internationally tradable product (e.g., oil), the supply cost is the opportunity cost of consuming the product domestically rather than selling it abroad plus any costs to deliver the energy to the consumer. Explicit subsidies also include direct support to producers, such as accelerated depreciation, but these are relatively small.

Implicit subsidies occur when the retail price fails to include external costs and/or there are preferential consumption tax rates on energy. External costs include contributions to climate change through greenhouse gas emissions, local health damages (primarily pre-mature deaths) through the release of harmful local pollutants like particulates, and traffic congestion and accident externalities associated with the use of road fuels. Getting energy prices right involves reflecting these adverse effects on society in prices and applying general consumption taxes to household fuels.

Source: IMF (2023).

There are thus related health, economic and other effects in the implicit subsidies (see Box 5.4).

Box 5.4: Health, Economic, and Ethical Issues About Fossil Fuel Subsidies

Subsidies are intended to protect consumers by keeping prices low, but they come at a high cost. Subsidies …

- encourage pollution—contributing to climate change and premature deaths from local air pollution
- have sizable fiscal costs—leading to higher taxes/borrowing or lower spending
- promote inefficient allocation of an economy's resources—hindering growth
- are not well targeted at the poor—mostly benefiting higher income households

Source: IMF (2023), as excerpted and reformatted by the author; also see Parry et al. (2021); Johnson (2009a, 2009b); Our World in Data (2022a, 2022b).

See Table 5.2 in Annex 5.3 for national subsidy amounts.

5.6 Mitigation Policies and Actions

The IEA (2022a, 2022b, 2022c, 2022d, 2022e) presents extensive analyses of what is needed for a "global pathway to net-zero emissions by 2050." The pathway includes many provisions for fossil fuels. For instance, "Net zero means a huge decline in the use of fossil fuels. They fall from almost four-fifths of total energy supply today to slightly over one-fifth by 2050" (IEA 2022c: 18). "Methane emissions from fossil fuel supply fall by 75% over the next ten years [approximately 2022–2032] as a result of a global concerted effort to deploy all available abatement measures and technologies" (IEA 2022d: 14).

Fossil fuel "key milestones" in the pathway include (IEA 2022c: 20: excerpted and compiled by the author):

2021: No new unabated coal plants approved for development
No new oil and gas fields approved for development
No new coal mines or mine extensions
2025: No new sales of fossil fuel boilers
2030: Phase-out of unabated coal in advanced economies
2040: Phase-out of all unabated coal and oil power plants

Such "bans and phaseouts" are one of the many kinds of policies in the "Climate Actions and Policies Measurement Framework (CAPMF)" developed by Nachtigall et al. (2022). The framework provides an extensive list of policies that can be used to reduce emissions. They are usefully organized according to economic sectors, such as those in subsequent chapters of the this book. In the present chapter, where the focus is on fossil fuels, the analysis draws upon fossil fuel examples from many sectors. Lists of relevant policies from the framework are also addressed in individual sectors in the next several chapters—electricity, transportation, industry, and buildings—where fossil fuels are also evident in sector-specific contexts.

In Box 5.5 some of the policies are explicitly and directly focused on fossil fuel production or fossil fuel use; others are indirect and implicitly include fossil fuels. The policies thus target Scope 1, 2 and 3 emissions, as defined in Sect. 5.3.

Box 5.5: Illustrative Mitigation Policies for Fossil Fuels

Bans and phase-outs are regulatory instruments that mandate the cessation of new (ban) or existing (phase out) infrastructure to extract fossil fuels. The CAPMF includes 4 policy variables. First, the due date (i.e. the year when the ban or phase out will be effective) for both bans and phase outs. Second, the legal status of both instruments. For these, the CAPMF distinguishes between (i) announcement, (ii) enshrined in law and (iii) achieved. The bans and phase-outs could include: ban and phase-out on fossil fuel extraction; ban and phase out on the construction of new unabated coal-fired power plants in the electricity sector; ban and phase out on fossil fuel heating systems in the buildings sector; ban public finance for fossil fuel infrastructure abroad in international programs; and ban governments' export credits for new unabated coal power in international programs.

Reform of fossil fuel producer support refers to limiting transfers or expenditures to producers of fossil fuels. Fossil fuel production encompasses the following activities along the supply chain such as exploration and extraction, bulk transportation and storage and refining and processing. The CAPMF includes one policy variable. This is the amount of fossil fuel producer support normalised by tax revenue.

Fuel excise taxes are levied on fossil fuels, implicitly putting a price on the carbon content of those fuels. The CAMPF includes up to 5 different policy variables for each sector, representing the most commonly used energy products. The taxes pertain to several sectors: electricity, transportation, buildings.

Policies to reduce fugitive methane emissions aim to reduce energy related methane emissions. The CAPMF includes one policy variable. This is the score for methane policies in the IEA methane tracker database (ranging from 0 to 7). The score includes policies such as restrictions on flaring or venting, as well as taxes or charges on emissions and mandatory technology use.

Air emission standards require coal power plants to observe specific emission limit values. The CAPMF includes four policy variables. They correspond to the emission limit values of four air pollutants: Nitrous oxide (NO_X), Sulfur oxides (SO_X), Particulate Matter (PM) and Sulphur.

Source: Excerpted and compiled, with italics added, by Thomas L. Brewer from Nachtigall et al. (2022: Annex Table B.1).

These numerous and diverse types of policies could be used widely and effectively, but of course, key features of actual implementations vary among countries. A tracking system with a data set based on the framework is available at the OECD (2023a, 2023b). The data set includes 128 policy variables for 52 countries, and it provides a "cross-country perspective of governments' climate actions and policies, building on the OECD Climate Actions and Policies Measurement Framework. It visualizes policy adoption and policy stringency (i.e. the degree to which climate actions and policies incentivize or enable GHG emissions mitigation) in a map … and across time ….” (OECD 2023a; also see OECD 2023b).

In short, it is clear that large-scale, stringent mitigation measures need to be adopted soon to reach zero emissions by 2050; but it is also clear that such measures are technologically feasible. Whether they are politically feasible is, of course, another matter.

Carbon Capture and Storage (CCS)

There is much interest in Carbon Capture and Storage (CCS) technology—which is particularly applicable to electric power plants using fossil fuels and to cement, steel and petrochemical facilities. There are three basic stages to the process Bandilla (2020): capturing CO_2 at a power plant or other facility, transportation of the CO_2 to a storage site, and injection into permanent underground storage. In chemical and structural terms, the stages of the CCS process include (adapted from Herzog, 2023; also see Herzog (2018):

Flue gas is diverted from smokestacks to a cooling tower
The flue gas is chemically bound to amines in an absorber
Carbon-free exhaust is vented into the air
A pure stream of CO_2 is created in a regenerator or stripper
A compressor converts the CO_2 gas into a fluid
The fluid is buried under ground or transported to be sold for a variety of uses, including enhanced oil well recovery

An early estimate IPCC (2005) was that CCS could contribute 15–55% to global cumulative CO_2 mitigation by 2100. However, there is controversy about the prospects for the technology. One issue is that the CCS systems themselves could use as much as 30% of the output of a power plant (Bandilla, 2020). There are also issues about leakage from transportation systems or storage sites IPCC (2005).

Nevertheless, Denmark and Norway announced that they had both issued licenses for CCS projects in their North Sea territories. These new projects are being undertaken in the aftermath of the collapse of projects in other EU countries and the UK before they got to the construction stage (Euractiv, 2015a, 2015b).

Denmark's new project is notable for being the world's first cross-national project, with captured CO_2 coming from Belgium and the Netherlands by ship and/or pipelines (Denmark, 2023). An affiliate of Total of France is an industry participant in the project, with other firms. The project is an important component of Denmark's national target of being CO_2 neutral by 2050; it hopes to become a “commercial hub” for storage of CO_2 in its North Sea territory, where there are many depleted oil and gas wells and geologically suitable areas below the seabed. The long-term plan is, therefore, to gain economically by providing carbon storage services for other countries while also achieving its own emission reduction plans. Norway's project includes BP as the industry participant and is located in geologically attractive Norwegian North Sea territory; like its Danish neighbor, it is planning to develop commercial storage services as well. In addition, there are depleted wells that provide ample capacity.

In sum, the Danish and Norwegian CCS projects are likely to establish them—and other EU institutional and corporate participants—as world leaders in the technologies, regulations, and economics of CCS.

5.7 Conclusion

Fossil fuels are core climate change problems in their extraction, processing and distribution stages, and in their consumption in the energy, transportation, and other sectors. Coal has traditionally been the most intensively polluting in its climate change effects as well as its health effects. However, the increasing quantities of oil and natural gas production and consumption also make them major climate change pollutants.

There are technological changes and governmental policies that could significantly reduce the emissions of fossil fuel industries, and there are ways to diminish the consumption of fossil fuels in many economic sectors. Despite action in those directions by some countries and sectors, the overall rate of progress in mitigating emissions has been far short of what is needed to achieve the Paris Agreement goals of limiting global temperatures to 1.5 °C or well below 2.0 °C. Three other international agreements in Annex 5.1 of this chapter may be helpful, though they are likely to yield only marginal effects compared with what is needed.

Annex 5.1: International Agreements on Fossil Fuels

Global Methane Pledge

Recognizing that, in order to ensure that the global community meets the Paris Agreement goal of keeping warming well below 2 °C, while pursuing efforts to limit warming to 1.5 °C, significant methane emission reductions must be achieved globally by 2030;

Recognizing that the short atmospheric lifetime of methane means that taking action now can rapidly reduce the rate of global warming and that readily available cost effective methane emission measures have the potential to avoid over 0.2 °C of warming by 2050 while yielding important co-benefits, including improving public health and agricultural productivity;

Recognizing that methane accounts for 17% of global greenhouse gas emissions from human activities, principally from the energy, agriculture, and waste sectors, and that the energy sector has the greatest potential for targeted mitigation by 2030;

Recognizing that the mitigation potential in different sectors varies between countries and regions, and that a majority of available targeted measures have low or negative cost;

Recognizing that, to keep 1.5 °C within reach, methane emission reductions must complement and supplement, not replace global action to reduce carbon dioxide emissions, including from the combustion of fossil fuels (coal, oil and natural gas) industrial processes, and the lands sector;

Recognizing that improvements to the transparency, accuracy, completeness, comparability, and consistency of methane emissions data assessed and validated in accordance with United Nations Framework Convention on Climate Change (UNFCCC) and Paris Agreement standards and Intergovernmental Panel on Climate Change (IPCC) good practice can promote more ambitious and credible action;

Recognizing that, while there are multiple useful international initiatives that address methane, there is a need for high-level political engagement in order to catalyze global methane action.

The Participants in the Global Methane Pledge:

Commit to work together in order to collectively reduce global anthropogenic methane emissions across all sectors by at least 30% below 2020 levels by 2030.

Commit to take comprehensive domestic actions to achieve that target, focusing on standards to achieve all feasible reductions in the energy and waste sectors and seeking abatement of agricultural emissions through technology innovation as well as incentives and partnerships with farmers.

Commit to moving towards using the highest tier IPCC good practice inventory methodologies, consistent with IPCC guidance, with particular focus on high emission sources, in order to quantify methane emissions; as well as working individually and cooperatively to continuously improve the accuracy, transparency, consistency, comparability, and completeness of national greenhouse gas inventory reporting under the UNFCCC and Paris Agreement, and to provide greater transparency in key sectors.

Commit to maintaining up-to-date, transparent, and publicly available information on our policies and commitments.

Commit to support existing international methane emission reduction initiatives, such as those of the Climate and Clean Air Coalition, the Global Methane Initiative, and the relevant work of the United Nations Environment Programme, including the International Methane Emissions Observatory, to advance technical and policy work that will serve to underpin Participants' domestic actions.

Welcome and encourage announcements of further parallel specific domestic actions by Participants and commitments taken by the private sector, development banks, financial institutions and philanthropy to support global methane abatement.

Resolve to review progress towards the target of the Global Methane Pledge on an annual basis until 2030 by means of a dedicated ministerial meeting.

Call on other states to join the Global Methane Pledge.

Source: Global Methane Pledge (2021a, 2021b).

Joint Declaration from Energy Importers and Exporters on Reducing Greenhouse Gas Emissions from Fossil Fuels

The United States, European Union, Japan, Canada, Norway, Singapore, and the United Kingdom are committed to taking rapid action to address the dual climate and energy security crises that the world faces.

We affirm the need to accelerate global transitions to clean energy, recognizing that reliance on unabated fossil fuels leaves us vulnerable to market volatility and geopolitical challenges.

We also recognize that under IPCC 1.5 °C-aligned scenarios, fossil fuel consumption will persist, at rapidly declining levels, as the global energy transition unfolds. As such, we emphasize that dramatically reducing methane, CO_2, and other greenhouse gas emissions across the fossil fuel energy value chain is a necessary complement to global energy decarbonization in order to limit warming to 1.5 °C.

We commit to taking immediate action to reduce the greenhouse gas emissions associated with fossil energy production and consumption, particularly to reduce methane emissions. We emphasize that reducing methane and other greenhouse gas emissions from the fossil energy sector enhances energy security by reducing avoidable routine flaring, venting, and leakage that wastes natural gas. We also note that these measures will also improve health outcomes by eliminating black carbon and other associated air pollutants.

We call on fossil energy importers to take steps to reduce the methane emissions associated with their energy consumption, which can spur emissions reductions across the value chain. We also call on fossil energy producers to implement projects and supporting policies and measures to achieve emissions reductions across fossil energy operations.

We call for global action to reduce methane emissions in the fossil energy sector to the fullest extent practicable, with the aim to reduce warming by 0.1 °C by midcentury, consistent with International Energy Agency findings of the near-term warming reduction effects of fully deploying technically feasible mitigation in this sector.

We reaffirm the call to action under the Global Methane Pledge to reduce collective anthropogenic methane emissions by at least 30% from 2020 levels by 2030 as an essential strategy to reduce warming in the near term and keep a 1.5 °C limit on temperature rise within reach. We recognize that the fossil energy sector must lead in rapid methane mitigation given the abundance of technically feasible and cost-effective mitigation measures available in the fossil energy sector, as called for in the Global Methane Pledge Energy Pathway.

Recognizing the urgency of reducing emissions from fossil energy value chains, we commit to working towards the creation of an international market for fossil energy that minimizes flaring, methane, and CO_2 emissions across the value chain to the fullest extent practicable, as we also work to phase down fossil fuel consumption. We support the development of frameworks or standards for fossil energy suppliers to provide accurate, transparent, and reliable information to purchasers about the methane and CO_2 emissions associated with their value chains.

We will support domestic and international action to achieve emissions reductions across the fossil energy value chain, such as:

- **Adopting policies and measures to achieve rapid and sustained reductions in methane and CO_2 emissions across the fossil energy value chain.**

Adopting policies and measures to eliminate routine venting and flaring and to conduct regular leak detection and repair campaigns in upstream, midstream, and downstream oil and gas operations.

Adopting policies and measures to capture, utilize, or destroy methane in the coal sector to the fullest extent practicable, including through pre-mine drainage, coal mine methane destruction, and ventilation air methane destruction.

Putting in place measures to require or strongly incentivize reductions in greenhouse gas emissions associated with fossil energy imports.

- **Adopting policies and measures to support robust measurement; monitoring, reporting, and verification; and transparency for methane emissions data in the fossil energy sector.**

Adopting policies and measures to improve the accuracy of methane emissions data, and affirming the need to enhance greenhouse gas inventories, including through improving data availability and through direct measurements at source level for gas and oil, in view of moving towards highest tier IPCC methods for emissions quantification based on direct measurement, stochastic sampling, emissions factors, and other IPCC-approved approaches, and improving monitoring, reporting, and verification mechanisms as new data becomes available.

Supporting frameworks or standards to improve the accuracy, availability, and transparency of fossil energy methane emissions and emissions intensity data at the cargo, portfolio, jurisdiction, and country level, including consideration of accepted protocols such as the Oil and Gas Methane Partnership 2.0 (OGMP2.0) standard and tools such as independent verification that can support robust data collection and reporting.

Supporting international efforts to improve methane emissions measurement; monitoring, reporting, and verification; and transparency, including through partnership with the UNEP International Methane Emissions Observatory and other multilateral partners.

Improving data quality on fossil energy methane, including for abandoned wells and mines, non-commercial operations, or retired infrastructure.

- **Strengthening coalitions to reduce methane and CO_2 emissions in value chains of internationally traded fossil fuels.**

Engaging public, municipal, and private sector fossil energy producers and purchasers to leverage contracts and other instruments, as appropriate, to improve methane and CO_2 emissions performance from traded fossil energy resources, including efforts to decrease the methane and other greenhouse gas intensity per unit of energy delivered.

Encouraging companies' participation in the Oil and Gas Methane Partnership 2.0 (OGMP2.0) standard.

- **Mobilizing technical assistance and financing for methane and CO_2 mitigation in the fossil energy sector.**

Enhancing the provision of technical assistance and investment for methane and CO_2 mitigation along the fossil energy value chain.

Developing financial tools and aligning financial standards to support methane and CO_2 mitigation in the fossil energy sector.

Source: US Department of State (2023).

Global Coal to Clean Power Transition Statement

We, the undersigned, noting that coal power generation is the single biggest cause of global temperature increases, recognise the imperative to urgently scale-up the deployment of clean power to accelerate the energy transition.

We commit to work together to make clean power the most affordable and accessible option globally, with ensuing economic and health benefits as we build back better from the COVID pandemic.

Our shared vision is to accelerate a transition away from unabated coal power generation, as is essential to meet our shared goals under the Paris Agreement, in a way that benefits workers and communities and ensures access to affordable, reliable, sustainable and modern energy for all by 2030 (SDG7).

'Unabated' coal power generation is described by the G7 and the IEA as referring to the use of coal power that is not mitigated with technologies to reduce carbon dioxide emissions, such as Carbon Capture Utilisation and Storage (CCUS). You can find out more in this G7 press release (July 2021) and on page 193 of the IEA 'Net Zero by 2050' report.

We commit to the following actions to drive this global transition forward, and we encourage others to make similar commitments:

1. To rapidly scale up our deployment of clean power generation and energy efficiency measures in our economies, and to support other countries to do the same, recognising the leadership shown by countries making ambitious commitments, including through support from the Energy Transition Council;

2. To rapidly scale up technologies and policies in this decade to achieve a transition away from unabated coal power generation in the 2030s (or as soon as possible thereafter) for major economies and in the 2040s (or as soon as possible thereafter) globally, consistent with our climate targets and the Paris Agreement, recognising the leadership shown by countries making ambitious commitments, including through the Powering Past Coal Alliance;
3. To cease issuance of new permits for new unabated coal-fired power generation projects ('New' coal-fired power generation projects are defined as coal-fired power generation projects that have not yet reached financial close), cease new construction of unabated coal-fired power generation projects and to end new direct government support for unabated international coal-fired power generation, recognising the leadership of countries making ambitious commitments, including through the No New Coal Power Compact;
4. To strengthen our domestic and international efforts to provide a robust framework of financial, technical, and social support to affected workers, sectors and communities to make a just and inclusive transition away from unabated coal power in a way that benefits them, and expands access to clean energy for all, recognising the leadership of countries endorsing the COP26 Just Transition Declaration.

We recognize that countries, workers, and communities in the developing world require support to transition from coal and realise a sustainable and economically inclusive energy future, and that international co-operation will be needed to provide such support.

We recognise that while significant progress has been made to realise our shared vision, our task is not yet complete, and we call on others to join us as we redouble our efforts to accelerate the global energy transition over the coming years.

[Signed by representatives of 46 national governments, 5 sub-national governments, and 26 other organizations.]

Source: UK COP 26 (2021).

Annex 5.2: Letters Exchanged by the British Royal Society and ExxonMobil in 2006

Introduction to letters
Royal Society and ExxonMobil
04 September 2006

The Society welcomes open debate, underpinned by sound science, on the subject of climate change. In September 2006, the Royal Society wrote to ExxonMobil to express concern that some of its corporate publications were presenting a misleading view of the scientific evidence about climate change and were overemphasising uncertainties about what we do and don't know. This letter followed a meeting which had taken place at the request of ExxonMobil where the Society raised concerns about Exxon's position on climate change and the company's funding of lobby groups that misrepresented the science. At the meeting ExxonMobil indicated that it intended to stop funding these organisations. The letter asked for clarification as to when the company would carry out this pledge. Although we have exchanged further letters with ExxonMobil, it has still not addressed this issue.

Source: Royal Society and ExxonMobil. https://royalsociety.org/topics-policy/publications/2006/royal-society-exxonmobil/https://royalsociety.org/topics-policy/publications/2006/royal-society-exxonmobil/ Accessed 11 May 2023.

Letter from the Royal Society to ExxonMobil, 4 September 2006. https://royalsociety.org/-/media/Royal_Society_Content/policy/publications/2006/8257.pdf Accessed 11 May 2023.

Letter from ExxonMobil to the Royal Society, 25 September 2006. https://royalsociety.org/-/media/Royal_Society_Content/policy/publications/2006/8257-Reply.pdf Accessed 11 May 2023.

Annex 5.3: Fossil Fuel Subsidy Levels

Table 5.2 Countries with Relatively Large Fossil Fuel Subsidies (Top 20 in 2020)

(a) Per capita (USD)	
Singapore	$1434.91
Kuwait	$567.27
United Arab Emirates	$566.66
Lebanon	$510.05
Saudi Arabia	$488.41
Libya	$480.29
Ireland	$433.57
Aruba	$424.17
Iran	$352.85
Suriname	$335.11
Bulgaria	$329.60
Belgium	$297.14
Australia	$286.16
Finland	$269.05
Denmark	$253.92
Turkmenistan	$250.56
Trinidad and Tobago	$229.14
Kazakhstan	$227.27

(continued)

Table 5.2 (continued)

Greece	$208.45
Algeria	$191.48
(b) Percent of GDP (%)	
Libya	16.65
Lebanon	9.92
Iran	6.74
Tajikistan	5.68
Palestine	5.07
Algeria	4.99
Venezuela	4.86
Turkmenistan	4.84
Kyrgyzstan	4.73
Suriname	4.45
Bulgaria	4.17
Uzbekistan	3.70
Mauritania	3.11
Ukraine	2.88
Timor	2.79
Northern Africa (UN)	2.66
Saudi Arabia	2.61
Iraq	2.59
Singapore	2.54
Kuwait	2.32

Source Excerpted and compiled by Thomas L. Brewer from Roser (2021)

Questions to Ponder

1. What are the similarities and differences in the emission profiles of oil, natural gas, and coal?
2. Which fossil fuel makes the most contribution to global warming? Why?
3. Assess natural gas as a "transitional" fuel to a more climate-friendly future. What are its advantages and disadvantages compared with oil and coal?
4. How do you think the "externalities" of coal should be addressed by government policies, if at all? Why? How?
5. Update a portion of a data set of interest to you. What has changed since then? Why? What are the consequences of the change(s)?

References

Bandilla, K. W. (2020). Carbon capture and storage. In *Future energy* (3rd ed.). Science Direct. https://www.sciencedirect.com/topics/earth-and-planetary-sciences/carbon-capture-and-storage. Accessed May 15, 2023.

Brower, D. (2023). *BP insists it is not slowing green transition to cash in on high oil prices.* https://www.ft.com/content/02facf98-e7c3-4973-beda-b1cc6e125d54. Accessed April 25, 2023.

Browne, J. (1997). *Climate change speech at Stanford University.* https://s3.documentcloud.org/documents/2623268/bp-john-browne-stanford-1997-climate-change-speech.pdf. Accessed June 20, 2022.

Carrington, D. (2023). *Turkmenistan moves towards plugging massive methane leaks.* https://www.theguardian.com/environment/2023/jun/13/turkmenistan-moves-towards-plugging-massive-methane-leaks. Accessed June 14, 2023.

Euractiv. (2015a). *UK cancels pioneering carbon capture and storage competition.* https://www.euractiv.com/?s=uk+cancels+pioneering. Accessed May 16, 2023.

Euractiv. (2015b). *Denmark inaugurates world's first cross-border CO_2 storage site.* https://www.euractiv.com/?s=denmark+inaugurates. Accessed May 16, 2023.

Fossil Fuel Subsidy Tracker. (2023). https://fossilfuelsubsidytracker.org/. Accessed May 14, 2023.

Global Carbon Project. (2023). *Global carbon project.* https://www.globalcarbonproject.org/index.htm. Accessed April 25, 2023.

Global Methane Pledge. (2021a). *About the global methane pledge.* https://mgp2020.wpengine.com/methane-policy-toolkit/. Accessed July 29, 2022.

Global Methane Pledge. (2021b). *Methane policy toolkit.* https://www.globalmethanepledge.org/. Accessed July 29, 2022.

Greenhouse Gas Protocol. (2023). *Standards.* https://ghgprotocol.org/about-us. Accessed April 22, 2023.

Hall, S. (2015). Exxon knew about climate change almost 40 years ago. *Scientific American.* https://www.scientificamerican.com/article/exxon-knew-about-climate-change-almost-40-years-ago/. Accessed August 23, 2022.

Herzog, H. (2018). *Carbon capture.* MIT Press.

Herzog, H. (2023). Climate portal. In *Carbon capture.* https://climate.mit.edu/explainers/carbon-capture. Accessed May 15, 2023.

Intergovernmental Panel on Climate Change (IPCC). (2005). *Carbon dioxide capture and storage.* Cambridge and New York: Cambridge University Press.

Intergovernmental Panel on Climate Change (IPCC). (2022a). *Climate change 2022: Mitigation of climate change. Annex III: scenarios and modelling methods.* The Working Group III Contribution to the Sixth Assessment Report. https://www.ipcc.ch/report/sixth-assessment-report-working-group-3/. Accessed August 11, 2022.

Intergovernmental Panel on Climate Change (IPCC). (2022b). *Climate change 2022: Mitigation of climate change.* Contribution of Working Group III to the Sixth Assessment Report of the Intergovernmental Panel on Climate Change. https://www.ipcc.ch/report/sixth-assessment-report-working-group-3/. Accessed August 11, 2022.

IPCC. (2023). *Synthesis report of the IPCC sixth assessment report (AR6), summary for policymakers.* https://report.ipcc.ch/ar6syr/pdf/IPCC_AR6_SYR_SPM.pdf. Accessed March 21, 2023.

International Energy Agency (IEA). (2022a). *Coal in net zero transitions.* https://www.iea.org/reports/coal-in-net-zero-transitions. Accessed January 12, 2023.

International Energy Agency (IEA). (2022b). *Coal 2022*. https://www.iea.org/reports/coal-2022. Accessed January 12, 2023.

International Energy Agency (IEA). (2022c). *Global methane tracker. Global methane emissions from the energy sector over time, 2000–2021*. https://www.iea.org/reports/global-methane-tracker-2022/estimating-methane-emissions. Accessed January 9, 2023.

International Energy Agency (IEA). (2022d). *Curtailing methane. Emissions from fossil fuel operations. Pathways to a 75% cut by 2030*. https://www.iea.org/reports/curtailing-methane-emissions-from-fossil-fuel-operations/executive-summary. Accessed April 26, 2023.

International Energy Agency (IEA). (2022e). *The growing evidence base and the international methane emissions observatory*. https://www.iea.org/reports/global-methane-tracker-2022/estimating-methane-emissions. Accessed January 9, 2023.

International Energy Agency (IEA). (2023a). Atlas of energy. http://energyatlas.iea.org/#!/tellmap/1378539487. Accessed March 31, 2023.

International Energy Agency (IEA). (2023b). *Methane tracker*. IEA. https://www.iea.org/data-and-statistics/data-tools/methane-tracker. Accessed March 31, 2023.

International Energy Agency (IEA). (2023c). *World energy outlook 2022*. https://www.iea.org/reports/world-energy-outlook-2022. Accessed March 31, 2023.

International Monetary Fund (IMF). (2023). *Climate change: Fossil fuel subsidies*. https://www.imf.org/en/Topics/climate-change/energy-subsidies. Accessed April 27, 2023.

Irakulis-Loitxate, I. (2022). Satellites detect abatable super-emissions in one of the world's largest methane hotspot regions. *Environmental Science and Technology*, *56*(4), 2143–2152. https://doi.org/10.1021/acs.est.1c04873

Johnson, J. (2009a). Energy's hidden cost. *Chemical Engineering and News, 19*, 10.

Johnson, J. (2009b). Fossil fuel costs. *Chemical Engineering and News, 26*, 5.

Mann, M. E. (2012) *The hockey stick and the climate wars: Dispatches from the front lines*. Columbia University Press.

Nachtigall, D., et al. (2022). The climate actions and policies measurement framework: A structured and harmonised climate policy database to monitor countries' mitigation action. In *OECD environment working papers, no. 203*. OECD Publishing. https://doi.org/10.1787/2caa60ce-en. Accessed April 17, 2023.

Orestes, N., & Conway, E. M. (2010). *Merchants of doubt*. Bloomsbury.

Organization for Economic Cooperation and Development (OECD). (2023a). Cross country analysis, in climate actions and policies measurement framework. https://oecd-main.shinyapps.io/climate-actions-and-policies/. Accessed April 17, 2023.

Organization for Economic Cooperation and Development (OECD). (2023b). *Policy instruments for the environment database*. https://www.oecd.org/environment/indicators-modelling-outlooks/policy-instruments-for-environment-database/. Accessed April 19, 2023.

Our World in Data. (2022a). *Excess mortality from fossil fuels*. https://ourworldindata.org/data-review-air-pollution-deaths. Accessed August 11, 2022.

Our World in Data. (2022b). *Population*. https://ourworldindata.org/explorers/population-and-demography?facet=none&Metric=Population&Sex=Both+sexes&Age+group=Total&Projection+Scenario=None&country=CHN~IND~USA~Europe~DEU~ITA. Accessed August 11, 2022.

Parry, W. H., Black, S., & Vernon, N. (2021). *Still not getting energy prices right: A global and country update of fossil fuel subsidies*. IMF Working Paper No. 2021/236. https://www.elibrary.imf.org/view/journals/001/2021/236/001.2021.issue-236-en.xml. Accessed April 27, 2023.

Reuters. (2023). *Denmark awards first CO_2 storage licenses in the North Sea*. https://www.euractiv.com/?s=denmark+inaugurates. Accessed May 16, 2023.

Ricketts, E. (2023). Supreme court declines to hear appeals from fossil fuel companies in climate change lawsuits. https://insideclimatenews.org/news/25042023/supreme-court-fossil-fuel-climate-change-lawsuits/. Accessed April 25, 2023.

Ritchie, H., & Roser, M. (2020). *CO_2 emissions by fuel*. OurWorldInData.org. https://ourworldindata.org/co2-and-greenhouse-gas-emissions. Accessed April 25, 2023.

Ritchie, H., Roser, M., & Rosado, P. (2020). *CO_2 and greenhouse gas emissions*. OurWorldInData.org. https://ourworldindata.org/co2-and-greenhouse-gas-emissions. Accessed April 25, 2023.

Roser, M. (2021). *Fossil fuel subsidies*. OurWorldInData.org. https://ourworldindata.org/fossil-fuel-subsidies. Accessed October 11, 2023.

Supran, G., & Oreskes, N. (2017). Assessing ExxonMobil's climate change communications (1977–2014). *Environmental Research Letters, 12*, 084019. https://doi.org/10.1088/1748-9326/aa815f/meta. Accessed April 26, 2023.

Supran, G., & Oreskes, N. (2020). Addendum to Assessing ExxonMobil's climate change communications (1977–2014). *Environmental Research Letters, 12*, 084019 (*Environmental Research Letters, 15*, 119401). https://doi.org/10.1088/1748-9326/ab89d5. Accessed May 11, 2023.

Supran, G., Rahmsdorf, S., & Oreskes, N. (2023). Assessing ExxonMobil's global warming projections. *Science, 379*, 6628. https://doi.org/10.1126/science.abk0063. Accessed April 26, 2023.

The Royal Society. (2006a). *The royal society and Exxonmobil*. https://royalsociety.org/topics-policy/publications/2006/royal-society-exxonmobil/. Accessed April 26, 2023.

The Royal Society. (2006b). *Letter from The Royal Society to ExxonMobil*. https://royalsociety.org/-/media/Royal_Society_Content/policy/publications/2006/8257.pdf. Accessed April 26, 2023.

The Royal Society. (2006c). *Letter to The Royal Society from ExxonMobil*. https://royalsociety.org/-/media/Royal_Society_Content/policy/publications/2006/8257-Reply.pdf. Accessed April 26, 2023.

UK COP 26. (2021). *Global coal to clean power transition statement*. https://ukcop26.org/global-coal-to-clean-power-transition-statement/. Accessed May 11, 2023.

United Nations Environment Programme and Climate and Clean Air Coalition. (2021a). *Global methane assessment: Benefits and costs of mitigating methane emissions*. United Nations Environment Programme. https://www.unep.org/resources/report/eye-methane-international-methane-emissions-observatory-2021-report. Accessed July 29, 2022.

United Nations Environment Programme and Climate and Clean Air Coalition. (2021b). *An eye on methane: international methane emissions observatory, 2021 report*. https://www.unep.org/resources/report/eye-methane-international-methane-emissions-observatory-2021-report. Accessed July 29, 2022.

US Department of State. (2023). *Joint declaration from energy importers and exporters on reducing greenhouse gas emissions from fossil fuels*. https://www.state.gov/joint-declaration-from-energy-importers-and-exporters-on-reducing-greenhouse-gas-emissions-from-fossil-fuels/. Accessed May 10, 2023.

Resources for Tracking Future Developments

Carbon Tracker. https://carbontracker.org
Fossil Fuel Subsidy Tacker. https://fossilfuelsubsidytracker.org
International Energy Agency (IEA). https://www.iea.org
Organization for Economic Cooperation and Development (OECD). https://www.oecd-ilibrary.org/environment/environment-at-a-glance-indicators_ac4b8b89-en
Our World in Data. https://ourworldindata.org

6 Electric Power

> Energy production and use is the single biggest contributor to global warming, accounting for roughly two-thirds of human-induced greenhouse gas emissions.
>
> United Nations Environment Programme (UNEP, 2023)

> Coal is both the largest source of electricity generation and the largest single source of CO_2 emissions, creating a unique challenge in transitioning to low-carbon energy systems.
>
> International Energy Agency (IEA, 2022a)

6.1 Introduction

The central issues in the electric power sector are: the phasing out of fossil fuels, the expansion of wind and solar capacities, and the development of smaller nuclear power reactors. Each of these poses a distinctive set of issues. For fossil fuels, the issues are the speed and scale of the phase out, the socio-economic impacts of the phase out, and of course related political conflicts and government policies. For solar and wind, the issues are mostly about economics, especially prices, technological advancements, and government subsidies and regulations. For nuclear power, the development of smaller reactors based on a new technology involves economic and regulatory issues as well as technological issues.

T. Brewer, *Climate Change*, https://doi.org/10.1007/978-3-031-42906-4_6

Section 6.2 of the chapter presents basic data about the patterns and trends in the sector's emissions. The replacement of fossil fuels by wind, solar and hydro technologies is discussed in Sect. 6.3. Mitigation options are the subject of Sect. 6.4. Nuclear options are considered in Sect. 6.5. The chapter summary is in Sect. 6.6. Annex 6.1 contains a statement from the multi-agency UN-Energy program with its "deliverables" for 2025.

6.2 Emissions

The electric power sector is the source of about one-third of the world's total carbon dioxide emissions (IEA, 2022a, 2022b), with an increasing share in some countries and decreasing share in others (IEA, 2021). In addition, methane emissions from power plants are major short-term climate pollutants (IEA, 2023). Table 6.1 indicates the CO_2-equivalent emissions per kilowatt hour of the various kinds of electric power plants. The estimates in the table are the 50th percentiles that take into account the variations in time and location of the plants (IRENA, 2022; Moomaw et al. 2011).

Table 6.1 Comparative carbon intensities of electricity generation technologies

Technology	g CO_2eq/kwh[a]
Fossil fuels	
Coal	1001
Oil	840
Natural gas	469
Other	
Hydro	4
Solar-photovoltaic	46
Solar-concentrated	22
Bio	18
Geothermal	45
Ocean	8
Nuclear	16

Sources Excerpted and reformatted by the author from IRENA (2022), which is based on data in Moomaw et al. (2011)

[a] These are the 50th percentile estimates in IRENA (2022) based on the literature review in Moomaw et al. (2011)

Depending on the mix of fuels, there are significant variations among countries in the relative carbon intensities of their electricity production. Countries with large domestic and/or imported coal supplies have relative high carbon intensities, while those with hydro or nuclear capacities have relatively low intensities. These and other variations are reflected in the data of Table 6.2.

The big emitters (China, India, and the US) in terms of their total carbon emissions have relatively carbon intensive electricity production sectors. The situation in Europe is different. The EU-27 is much lower, but there are major differences among the member counties. Poland is very high at more than twice the EU-27 level as a group, and Germany is also above the collective EU-27 level. France, in contrast, is not only the lowest among the EU-27, it is among the lowest in the world. Elsewhere in Europe, Switzerland is even lower. At the other end of the scale, globally, Australia and Saudi Arabia are among the highest in the world.

The cases of the countries with extremely low and extremely high levels are readily explained. Switzerland has an extensive hydro-based electric power capacity, and France has nearly three-fourths of its total capacity in nuclear. At the high end, China, India and Australia get between two-thirds and three fourths of their electricity from fossil fuels. Poland and Saudi Arabia are even higher, though the latter has plans for large-scale solar installations.

Table 6.2 Comparative carbon intensities of illustrative countries' electricity generation

Countries	g CO_2eq/kwh[a]
Carbon intensive emitters	
China	544
India	637
US	379
EU-27	278
Germany	386
Poland	635
Saudi Arabia	571
Australia	531
Low carbon intensity emitter	
Switzerland	47
World average	441

Source Our World in Data (2023)

Table 6.3 Renewable energy capacities

	2022 (GW)[a]	Percent change 5 years (2017–2022)	Percent of world total
Total	3372	54	100
Hydro	1393	10	41
Solar	1053	166	31
Wind	899	73	27

Source Computed from IRENA (2023a: 2, 6, 14, 21)
[a] Rounded to GW from MW in the source

6.3 Renewable Technologies

The differences in carbon emission rates among fossil fuels and the alternatives for producing electricity are thus evident. The increasing economic competitiveness and production capacities of solar and wind installations are also clear. Hydro is a special case because it is not growing much globally and is in fact decreasing in some countries (see Table 6.3).

As for government policies, the situation is more complex. Political resistance to phasing out coal is particularly strong in Germany and Poland in Europe, and in Australia, and some regions of the US. At the same time, there are many government mitigation policy options available.

6.4 Mitigation Options

"Avoided emissions" are estimates of the "emissions that have been avoided due to a country's uptake of renewables in a given year." The value varies "greatly depending on the non-renewable mix that has been replaced by renewables" (IRENA, 2023b: 1). Table 6.4 displays the amount of "avoided emissions" in 2020 for the four biggest national emitters in that year. Replacements of coal fired power plants accounted for more avoided emissions than other non-renewable technologies.

Table 6.4 "Avoided emissions" in "big emitter" countries from the use of renewables

Country	Million tonnes CO_2-eq. (2020)	Assumed percent from coal
China	1922	83
India	277.5	91
US	394.7	50
Germany	154.3	50

Sources IRENA (2023b)

The wide variety of mitigation options that are available are outlined in Box 6.1, as identified in the Climate Actions and Policies Measurements Framework (CAPMF).

Box 6.1: Mitigation Options for the Electricity Sector

An emissions trading scheme (ETS) or cap-and-trade is a market-based instrument that aims at controlling and reducing emissions in a cost-effective manner. The CAPMF includes two policy variables on ETS. First, the average permit price observed in each year in EUR per tCO_2e. Currencies other than EUR were converted to EUR using 2020 real exchange rates. Second, the coverage of GHG emissions differentiated by CO_2, CH4, N20 and all other GHG. Coverage of GHG gases is weighted by the contribution of each gas to global GHG emissions. The CAPMF also includes all sub-national ETSs. Currencies other than USD were converted to USD using 2020 real exchange rates.

Carbon taxes are levied on carbon emissions to reduce carbon emissions. The CAPMF includes one policy variable on carbon taxes in each sector. This is the nominal tax rate measured in USD per tCO_2e. Currencies other than USD were converted to USD using 2020 real exchange rates.

Feed in tariffs for solar PV and wind (FiT) are policy instruments that spur investments into renewable energy by offering fixed long-term contracts to renewable energy producers. The CAPMF includes four policy variables on FiT: First, the contract length of the FiT for both solar photovoltaic and wind power. Second, the price of the FiT for both technologies. The price of the FiT is normalised by the global levelised cost of electricity to account for falling technology costs, following the EPS 2022 update.

Renewable energy auctions are competitive tenders issued by the government to install a specific capacity of renewable capacity. The CAPMF includes four policy variables on auctions: First, the contract length of the auction for both solar photovoltaic and wind power. Second, the auction price, normalised by the global levelised cost of electricity to account for falling technology costs for both technologies. If multiple auctions are held in one year, the contract length and the price are calculated as the weighted average (by capacity) of those auctions.

Renewable energy portfolio standards (RPS) mandates electricity generators to cover a specific share of their output by renewables. In most cases, RPS allow trading of renewable energy certificates to comply with the standard at lower cost. The CAPMF includes one policy variable: the mandated percentage of renewable production on total production. The CAPMF also includes sub-national RPS. These are constructed as weighted averages (by total electricity generation) of sub-national RPS.

Ban on the construction of new and phase out of existing unabated coal power plants includes four policy variables. First, the due date (i.e. the year when the ban or phase out will be effective) for both bans and phase outs. Second, the legal status of both instruments. For these, the CAPMF distinguishes between (i) announcement, (ii) enshrined in law and (iii) achieved.

Air pollution standards for coal power plants require coal power plants to observe specific emission limit values. The CAPMF includes four policy variables. They correspond to the emission limit values of four air pollutants: Nitrous oxide (NOx), Sulfur oxides (SOx), Particulate Matter (PM) and Sulphur.

Planning for renewables expansion refers to integrated transmission and (renewable) generation planning in combination with resource data and siting. It is a key enabling condition to expand generation from renewable energy sources. The CAPMF includes one policy variable. This is the final score for planning for renewables in the RISE database to ensure consistency with the underlying data. This score is derived from a questionnaire, containing seven planning-related questions such as whether renewable energy is included in transmission planning or whether there are policies on resource data and siting.

Source: Excerpted and compiled, with italics added, by Thomas L. Brewer from Nachtigall et al. (2022: Annex Table B.1).

From this wide-ranging agenda of possibilities, there are many specific combinations of options that can be imagined and implemented. One list of a package of policies that could be developed to achieve net zero CO_2 energy systems has been summarized by the IPCC (2023: e.g. Sect. 4.5.1 and Fig. SPM.6; excerpted by the author) as involving:

> … a substantial reduction in overall fossil fuel use, minimal use of unabated fossil fuels, and use of carbon capture and storage in the remaining fossil fuel systems; electricity systems that emit no net CO; widespread electrification; alternative energy carriers in applications less amenable to electrification; energy conservation and efficiency; and greater integration across the energy system ….
>
> Large contributions to emissions reductions with costs less than USD 20 tCO2-eq come from solar and wind energy, energy efficiency improvements, and methane emissions reductions …. Energy generation diversification (e.g., via wind, solar, small scale hydropower) and demand side management (e.g., storage and energy efficiency improvements) can increase energy reliability and reduce vulnerabilities to climate change ….

Note that nuclear alternatives are not included in this package. This is partly because nuclear power technologies are typically not included in "renewables" lists. This is more than a semantic issue; it reflects a division of opinion among advocates of phasing out fossil fuels. The International Renewable Energy Agency (IRENA), in particular, does not include nuclear in its renewables list.

6.5 Nuclear Alternatives

There are, of course, many issues about nuclear energy in addition to its contributions to climate change mitigation—including safety, economics, waste disposal, and nuclear weapons proliferation. The focus here is on the technological and economic issues that are directly related to questions about its past and potential effects on climate change.

Nuclear's share of the world total electricity production declined from about 15% in 1985 to 10% in 2022 (Our World in Data, 2023). If the world total is disaggregated to the regional and country levels, the trends have been mixed. Several East European countries increased their nuclear shares, while Germany phased out its nuclear industry completely. In 2011, the German government announced that it would begin immediately to reduce its 17 reactors to eight, and to zero by the end of 2022. Because of the energy crisis in Europe at that time, the final three reactors were allowed to shut down a few months later in 2023. France, meanwhile, maintained one of the largest nuclear capacities in the world with nearly two-thirds of its electricity being produced by nuclear reactors. In Asia, meanwhile, China, India, Japan, and South Korea were all building new reactors for addition to their existing facilities. As of mid-2019, "developing economies" were adding about 37% to their existing nuclear capacities of 110 GW, while "advanced economies" were adding about 6% to their existing 312 GW (IEA, 2019).

During the period of growing nuclear capacity from 1971 to 2018, "Without nuclear power, emissions from electricity generation would have been almost 20% higher …" (IEA, 2019). However, the Chernobyl accident in Russia in 1986 and the Fukushima accident in Japan in 2011 led to reductions in nuclear reactor construction starts and the nuclear share of total electricity production in many countries.

As for the future, nuclear expansion options can be simply summarized as consisting of (a) extending the lifetime of existing large-scale reactors, (b) adding new large-scale reactors with similar technologies, and (c) developing and deploying smaller-scale reactors with new technologies. There are also two contraction options: (i) waiting for the end of the useful lives of existing reactors without replacing them, and (ii) shutting down existing reactors before the end of their useful lives.

The age of existing rectors is an important factor in projecting future capacities. In particular, the average age of the reactors in the US in the early 2020s is about 40 years old and about 35 years in Europe, while those in China are about ten years and in India about 20 years (IEA, 2019). The policy options are therefore different. In Europe and the US, the ageing fleets of reactors pose both problems and opportunities. The problematic challenge is whether to invest large amounts of money in large old reactors to extend their lifetimes—or take advantage of an opportunity to invest in an emerging but not yet ready for deployment new technology, which has many advantages. In contrast, for China, India, and some countries in Eastern Europe, where the expected lifetimes of their existing reactors are still on the order of several decades, it is quite a different decision.

The contribution of nuclear technology to the future of climate change emissions in the electric power industry, therefore, depends on a complex combination of technology, economics, and politics. Although this may not be a very satisfying summary, it is realistic in the early 2020s.

6.6 Conclusion

Solving the climate change crisis requires the reduction of the use of fossil fuels in the production of electricity. The issues are: how fast, at what scale, and with which technological alternatives. Solar and wind technologies have clearly become economically competitive as well as environmentally preferable to fossil fuels. The use of nuclear is more complex and more controversial. Therefore, whereas the specifics of the transition from fossil fuels to solar and wind largely depend on economic differences, the transition to nuclear alternatives will largely depend on political processes and further development of new technologies.

Annex 6.1: UN Energy Program 'Deliverables' for 2025

- 500 million more people with electricity access.
- 100% increase in modern renewables capacity globally and 100% renewables-based power targets established in 100 countries.
- 3% annual efficiency improvement in at least 50 countries across the world.
- 30 million jobs in renewable energy and energy efficiency.
- Redirect fossil fuel subsidies towards clean energy and end financing for new coal power plants.
- Annual global GHG emissions to be reduced at least by one third in 2025.
- Double annual clean energy investment globally (relative to the current level).
- Raise energy access investment to US$ 40 billion of which 50% is directed to the Least Developed Countries (LDCs).

Statement About the Opportunities

Energy can create transformational opportunities. Investing in clean, affordable and sustainable energy solutions will end energy poverty, spur innovation, grow multi-trillion-dollar markets, generate tens of millions of green jobs, develop sustainable cities and communities, and help create a just, equitable, net-zero future that leaves no one behind and makes peace with nature.

Source: UN-Energy (2022).

Questions to Ponder

1. What do you think are significant constraints—if any—in the transition from fossil fuels to solar and wind to produce electricity?
2. Which countries do you think are keys to the transition from fossil fuels to the alternatives? Why?
3. How do the new generation of nuclear reactors compare with previous generations? Do you think they will have a significant effect on reducing the carbon emissions of the electricity sector?
4. Which countries are leaders in the transition from fossil fuels to alternatives? Why are they leaders?
5. Do you think a phase out of fossil fuels in electricity production will be sufficient to avoid catastrophic consequences from climate change? Why or why not?

References

International Energy Agency (IEA). (2019). Nuclear power in a clean energy system. https://www.iea.org/reports/nuclear-power-in-a-clean-energy-system. Accessed May 31, 2023.

International Energy Agency (IEA). (2022a). Coal. https://www.iea.org/fuels-and-technologies/coal. Accessed July 11, 2022.

International Energy Agency (IEA). (2022b). Coal 2021: Analysis and forecast to 2024. https://www.iea.org/fuels-and-technologies/coal. Accessed July 11, 2022.

International Energy Agency (IEA). (2023). Global methane tracker 2023. https://www.iea.org/reports/global-methane-tracker-2023. Accessed May 30, 2023.

Intergovernmental Panel on Climate Change (IPCC). (2023). Synthesis report of the IPCC sixth assessment report (AR6), summary for policymakers. https://report.ipcc.ch/ar6syr/pdf/IPCC_AR6_SYR_SPM.pdf. Accessed March 21, 2023.

International Renewable Energy Agency (IRENA). (2021). Methodology: Avoided emissions calculator. https://www.irena.org. Accessed May 13, 2023.

International Renewable Energy Agency (IRENA). (2022). World Energy Transitions. https://www.irena.org/Digital-Report/World-Energy-Transitions-Outlook-2022. Accessed October 8, 2023.

International Renewable Energy Agency (IRENA). (2023a). Renewable energy capacity statistics. https://www.irena.org. Accessed May 13, 2023.

International Renewable Energy Agency (IRENA). (2023b). Avoided emissions calculator. https://www.irena.org/Data/View-data-by-topic/Climate-Change/Avoided-Emissions-Calculator. Accessed May 13, 2023.

Moomaw, W., et al. (2011). Annex II: Methodology. In *IPCC special report on renewable energy sources and climate change mitigation*. Cambridge University Press.

Nachtigall, D., et al. (2022). The climate actions and policies measurement framework: A structured and harmonised climate policy database to monitor countries' mitigation action. In *OECD environment working papers, no. 203*. https://doi.org/10.1787/2caa60ce-en. Accessed October 8, 2023.

Our World in Data. (2023). Energy: Key charts. https://ourworldindata.org/energy-key-charts. Accessed May 30, 2023.

UN-Energy. (2022). UN-energy plan of action. https://un-energy.org/wp-content/uploads/2022/05/UN-Energy-Plan-of-Action-towards-2025-2May2022.pdf. Accessed June 1, 2023.

United Nations Environment Programme (UNEP). (2023). Theme report on energy transition. https://www.un.org/sites/un2.un.org/files/2021/11/2021-twg_2.pdf. Accessed May 17, 2023.

Resources for Keeping Up with Developments in the Electric Power Sector

Intergovernmental Panel on Climate Change (IPCC). https://www.ipcc.ch/

International Atomic Energy Agency (IAEA). https://www.iaea.org/

International Energy Agency (IEA). https://www.iea.org/reports/world-energy-outlook

International Renewable Energy Agency (IRENA). https://www.irena.org/

Organization for Economic Cooperation and Development (OECD). https://data.oecd.org/energy/renewable-energy.htm

United Nations Environment Programme (UNEP). https://www.unep.org/explore-topics/energy

7 Transportation

> Transport's share of total national GHG emissions range from up to 30% in high income economies to less than 3% in LDCs.
>
> United Nations, Sustainable Transport Conference (2021)

7.1 Introduction

The transportation subsectors are highly heterogeneous in the technological, economic, and political issues that they pose for climate change problems and solutions. Nevertheless, the subsectors have a common central feature—namely that they are significant sources of carbon emissions from fossil fuels. The dominant issues about the sector and its subsectors are: whether, how, how extensively, and how quickly they are transitioning to alternative fuels.

This chapter considers these issues initially in Sect. 7.2, where the fundamental patterns and trends in carbon emissions are analyzed. Section 7.3 focusses on the industry technologies and government policies that can reduce the emissions in the individual subsectors. Section 7.4 summarizes the chapter.

7.2 Emission Patterns and Trends

An estimated 72% of transportation emissions in 2020 came from road vehicles, followed by 12% from maritime shipping, 9% from aviation, and 7% from rail and other sources. In 2021, transport emissions began to rise again, as the covid pandemic receded. The sector recovered about 44% of the decrease in CO_2 emissions from 2019 to 2020.

T. Brewer, *Climate Change*, https://doi.org/10.1007/978-3-031-42906-4_7

Table 7.1 Transportation CO_2 emissions (2021)

Mode	Percent of transportation
Motor vehicles	78
Trains	1
Aviation	9
Shipping	11

Sources Computed from IEA (2022)

Table 7.2 Seaborne trade and emissions (1990–2020)

Volume of emissions and trade	1990	2000	2010	2020	Percent change: 1990–2010	Percent change: 2010–2020
Emissions (Mt)	526	709	1135	1056	Increase: +116	Decrease: −7
Trade (Mt)	4008	5064	8231	11,019	Increase: +105	Increase: +34

Source Excerpted and reformatted from IMO (2023b)

In 2021, the emissions were as displayed in Table 7.1.

The dominance of *motor vehicles* is a persistent feature of the transportation sector. At the other extreme, *trains* have been the least carbon intensive mode. *International aviation* emissions during the covid pandemic reflected the ups and downs of the passenger loads; they were higher in 2021 than 2020 but not back to the higher pre-pandemic level in 2019 (Liu et al. 2023; IEA, 2022).

In *international shipping*, an important feature of the emissions of international cargo shipping is that the volumes of emissions in recent years have become less correlated with volumes of the cargo. Emissions peaked about 2010 and then declined, even though the amount of traffic continued to climb after 2010. The volume of emissions decreased by seven percent over the 2010–2020 decade as the amount of cargo increased by 34% (see Table 7.2).

The *decoupling* that is evident in the table is consistent with the conclusions of a study highlighted as follows (italics added, excerpts taken from CE Delft 2023):

- It is technically possible to reduce shipping emissions by 28–47% by 2030, relative to 2008 … This amounts to approximately 175–350 Mt CO2e … per annum.
- When introduced gradually from 2025, the measures could avoid cumulative emissions of 500–1000 Mt CO2e.
- About *half* of the emission reductions [would] result from lower speeds and other operational measures, a *quarter* from wind-assisted propulsion and other technical measures, and another *quarter* from using zero and near-zero-GHG fuels.
- Implementing these measures would increase shipping costs by 6–14% on average, relative to BAU.

Cruise ships pose many of the same issues as cargo ships. However, their black carbon (BC) emissions are especially problematic (Fleck, 2022). The annual BC emissions of cruise ships average 10 tonnes per ship, compared with 3.5 tonnes for container ships. Since there are many more container ships than cruise ships, the total BC contribution of container ships is much greater. Yet, an indication of the disproportionate amount of BC contributed by cruise ships is that they contribute 6% of the total global fleet of cargo and cruise ships, though they are only 1% of the number of ships.

Another feature of cruise ships is that they spend a disproportionate amount of time near coasts, including ports in large urban areas. Since much of the BC particulate matter emissions fall in the immediate region of the ships, there is a more intensive localized effect from the localized depositions of BC compared with the more globalized circulation of carbon dioxide and other airborne ghg gasses (Brewer, 2020a, 2020b, 2020c, 2020d, 2023). Finally, BC poses serious lung and cardiovascular health problems. In short, because BC emissions are especially problematic as health hazards in densely populated port cities, there are significant health co-benefits along with the climate change benefits of reducing cruise ships' emissions. For comparisons of cruise ships with airplanes, see Comer (2022).

7.3 Technologies and Policies

There are many technologies and policies that can have significant emission mitigation effects for each transportation mode. The periodically updated ratings of the Climate Action Tracker are useful summary overviews with supporting detailed analyses. A sample overview follows (Climate Action Tracker, 2022):

Transforming the global transportation system … will require three key shifts …

- First, travel must shift to or remain as active modes (including walking and bicycling) and shared public transport. For this shift, this report tracks short- and medium-distance mode shift via the share of kilometers traveled by passenger cars, the kilometers of urban rapid transit per 1 million inhabitants, and the kilometers of high-quality, safe urban bike lanes per 1000 inhabitants. Long-distance mode shift is not accounted for in this report due to data limitations and space constraints.
- Second, governments must phase out the internal combustion engine and move to zero-carbon road vehicles.
- Finally, the shipping and aviation systems must decarbonize through a combination of demand-reduction strategies and zero-carbon technologies.

Motor vehicles

The "Climate Actions and Policies Measurement Framework (CAPMF)" (Nachtigall et al. 2022) includes many policies that can be used to reduce emissions of

Table 7.3 Emission mitigation policies focused on motor vehicles in the "Climate Actions and Policies Measurement Framework (CAPMF)"

Policy	Number of countries where policies were adopted as of 2020[a]
Congestion charges	4
Emissions trading scheme	4
Carbon tax	18
Fossil fuels subsidies reform	34
Fossil fuels excise taxes	44
Minimum energy performance standards	40
Energy labels	40
Speed limits on motorways	45
Ban and phase out of passenger cars with ICE [Internal Combustion Engine]	14

Source Excerpted and compiled by the author from (Table 3.1)
[a]Number of countries that adopted the policy, among the 52 in the International Programme for Action on Climate (IPAC)

motor vehicles. Table 7.3 lists policies that can reduce such emissions, and it also includes the number of countries with such policies as of 2020.

Several of these kinds of policies are further described in Box 7.1.

Box 7.1: Illustrative Mitigation Policies for Motor Vehicles

Congestion charges are a daily levy imposed on drivers who chose to drive within a given area of a city. The CAPMF includes one policy variable. This is the price at peak hour of a city's congestion charges.

An emissions trading scheme (ETS) or cap-and-trade is a market-based instrument that aims at controlling and reducing emissions in a cost-effective manner. The CAPMF includes two policy variables on ETS. First, the average permit price observed in each year in EUR per tCO2e. Currencies other than EUR were converted to EUR using 2020 real exchange rates. Second, the coverage of GHG emissions differentiated by CO_2, CH_4, N_20 and all other GHGs. Coverage of GHG gases is weighted by the contribution of each gas to global GHG emissions. The CAPMF also includes all sub-national ETSs. Currencies other than USD were converted to USD using 2020 real exchange rates.

Carbon taxes are levied on carbon emissions to reduce carbon emissions. The CAPMF includes one policy variable on carbon taxes in each sector. This is the nominal tax rate measured in USD per tCO2e. Currencies other than USD were converted to USD using 2020 real exchange rates.

Ban on the sales of new and phase out of conventional passenger cars are regulatory instruments that mandate the cessation of the purchase (ban) or the usage (phase out) of certain activities, here passenger cars with internal combustion engines. The CAPMF includes 4 policy variables. First, the due date (i.e. the year when the ban or phase out will be effective) for both bans and phase outs. Second, the legal status of both instruments. For these, the CAPMF distinguishes between (i) announcement, (ii) enshrined in law and (iii) achieved.

Source: Excerpted and compiled, with italics added, by the author from Nachtigall et al. (2022: Annex Table B.1).

International Aviation

The Carbon Offsetting and Reduction Scheme for International Aviation (CORSIA) is the core agreement to date at the International Civil Aviation Organization (ICAO, 2023b). It is a market-based measure with the goal of net zero emissions *growth—not net zero emissions.* It has three phases: a pilot phase (2021–2023), a first phase (2024–2026), and a second phase (2027–2035). Participation is voluntary in the pilot phase and first phase. Thus, mandatory participation is delayed until 2027. An intention to participate had been declared by 115 countries as of the beginning of 2023.

CORSIA has been assessed by Climate Action Tracker (2022) as follows:

> We rate the target of carbon neutral growth from 2020 as "Critically insufficient." Under the CAT's rating methodology, the upper end of the target range in 2030 alone would be rated "Highly insufficient." Because the international aviation sector plans to rely on emission units and alternative fuels that are unlikely to deliver sufficient real emission reductions, we downgrade the rating to "Critically insufficient."
>
> A further shortcoming of ICAO's approach is it does not address indirect GHG emissions and impacts– such as NO_X and contrail cirrus. These emissions and impacts are responsible for an estimated two thirds of aviation's effective radiative forcing impact (Lee et al. 2021).

Indeed, airplane contrails have become a focus of attention among critics of the ICAO approach to climate change issues, and Box 7.2 considers the issue in detail.

Box 7.2: Airplane contrails

It has been established by climate science research for more than two decades that an airplane's white vapor trails called "contrails" are significant climate change forcing agents (Fahey, 2015; Irfan, 2011). In fact, Burkhardt and Karcher (2011) concluded more than a decade ago that "net radiative forcing due to contrail cirrus remains the largest single radiative-forcing component associated with aviation."

A study by the European Union Aviation Safety Agency (EASA) (2020) for the European Commission found that climate pollutants other than

CO_2 contribute about twice as much as CO_2 to airplanes' global warming impact. The non-CO_2 emissions include black carbon and methane as well as nitrogen oxides (which have indirect effects via effects on ozone and methane).

It should be noted that a 100-year Global Warming Potential (GWP) was used to compute the CO_2-equivalent, as it is widely used in the EU, including in the Emission Trading System (ETS). If the commonly used 20-year GWP were applied, the contributions of black carbon and methane would be much greater since they are both highly potent short-term forcers.

There are, of course, technical complexities and uncertainties needing further analysis, but there is also a basis for a wide range of technological and operational options for non-CO_2 emissions mitigation. For instance, "Reducing soot particle emissions … by means of sustainable low carbon footprint aviation fuels, would be a 'win–win' situation for improving air quality [and thus fewer deaths] and reducing contrail cirrus impact on climate" (EASA, 2020). This and many other changes in current industry technologies and operations could be the focus of EU policies, such as those in Table 7.4. See the extensive discussion of the policies in EASA (2020) for further information.

Table 7.4 Potential policies to reduce aviation non-CO_2 emissions

Policy type		Main non-CO_2 emission mitigation effects
Financial	Add NO_x charge to each flight	NO_x emissions
	Include NO_x emissions in EU Emission Trading System (ETS)	NO_x emissions
Fuel	Reduce aromatic limit in fuel standards	Soot [black carbon] emissions
	Mandate Sustainable Aviation Fuels (SAFs)	Soot [black carbon] emissions
Air Traffic Management	Avoid flights in ice-supersaturated areas [which induce bigger contrails]	Contrail size
	Impose climate charge to each flight	All non-CO_2 emissions [and potentially CO_2]

Source Adapted from EASA (2020: 13)

At the International Civil Aviation Organization (ICAO), contrails have received some attention. In the text of the Chicago Convention that created ICAO in 1944, In Annex 16—Environmental Protection—Volume II—Aircraft Engine Emissions, soot [black carbon] emissions are mentioned (ICAO, 2023a, 2023b). An ICAO (2016) "White Paper on Climate Change Aviation Impacts On Climate: State Of The Science" presented "a summary of recent progress in the state of the science especially related to contrails and induced cloudiness, contrail avoidance, and aerosol and NOx effects." However, there has not been a sustained focus in ICAO on possible actions to address contrail issues.

Sources: In addition to the items cited in the paragraphs above, see IPCC (2013), Lee et al. (2021), Teoh et al. (2020), Transport & Environment (2020).

Annex 7.1 of this chapter includes a chronological account of ICAO's attention to climate change.

There are many kinds of governmental policies and industry practices that could reduce aviation emissions in the short-term, medium-term, and long-term. They include the measures indicated in Table 7.4—namely finances, fuels, and air traffic management.

International shipping

In addition to several fuel efficiency measures developed over a long period, the IMO approach to climate change is now embodied in its "Strategy to Reduce GHG Emissions in International Shipping," which was agreed and made public in 2018. Its objective is to reduce carbon emissions from international shipping by 40% by 2030 and by 70% by 2050, compared with 2008. It was scheduled for review in 2023. This chapter's Annex 7.2 contains a chronology of developments concerning climate change at the IMO.

Climate Action Tracker (2021) has evaluated IMO's record as follows: "The IMO continues to move at a snail's pace.... The [shipping] sector's emissions will continue to grow through to 2050 unless there is further policy action."

In mid-2023, there was a potentially significant agreement—technically a climate roadmap of the Marine Environment Protection Committee (The Maritime Executive, 2023). The agreement includes emission reduction targets compared to 2008 of 20% by 2030, 70% by 2040, and net-zero emissions "by or around, i.e. close to, 2050." The ambiguity about 2050 is one limitation. Another is that the targets are not binding.

7.4 Conclusion

The transportation sector remains a major source of greenhouse gas and particulate pollutants. The overall recent levels and trends are not adequate to prevent potentially catastrophic consequences worldwide. All subsectors are under much pressure to adopt new technologies and new operational practices as the transitional period toward more sustainable industries progresses.

Just as the patterns and trends in emissions have varied considerably among industry subsectors, so too have the speed and scale of the uptake of new technologies and the adoption of new industry practices. To date, the motor vehicle industry has been a leader among the subsectors in technological change with the adoption of electric power to replace fossil fuels. The aviation sector has been a laggard. Even within individual subsectors, inside national boundaries, there have been important local governmental initiatives; this has been true, for instance, in airports and harbors. Overall, though, the sector's trends do not bode well for the future.

Annex 7.1: Chronology of Climate Change Issues at the International Civil Aviation Organization (ICAO)

1944 ICAO created to promote safe and efficient civil aviation [no mention of environmental issues].

1983 Established Committee on Aviation Environmental Protection (CAEP).

2013 Assembly set an *aspirational goal* for international aviation of *carbon neutral growth* from 2020 levels. Carbon neutral *growth* means that net CO_2 emissions from international aviation *remain constant compared to the baseline*—i.e. set at average emissions level in 2019–2020 [Due to COVID-19, aviation emissions in 2020 were substantially lower than anticipated, which would have meant the baseline for carbon neutral growth would have been lower than anticipated and the target more ambitious. The Council thus changed the reference year for CORSIA's pilot phase to 2019.]

2016 *Report* on Carbon Offsetting and Reduction Scheme for International Aviation (CORSIA).

2019 *Agreement* to establish Carbon Offsetting and Reduction Scheme for International Aviation (CORSIA), which is a global market-based system. A voluntary pilot phase was scheduled to begin in 2021.

2020 ICAO Council *changed* the reference year for CORSIA's pilot phase to 2019, before the COVID pandemic to reduce the targeted emission levels.

2021 CORSIA *pilot* phase began, with voluntary participation.

2023 As of beginning of the year, 115 members had indicated their *intention* to participate in CORSIA. At end of the year, voluntary pilot phase ended.

2024 The scheduled beginning of the "first" phase of CORSIA, following the end of the *pilot* phase in 2023.

Sources: ICAO (2016, 2023a, 2023b).

Annex 7.2: Chronology of Climate Change Issues at the International Maritime Organization (IMO)

1948 International conference in Geneva adopted a convention formally establishing IMO (The original name was the Inter-Governmental Maritime Consultative Organization, or IMCO, but the name was changed in 1982 to IMO.)

1958 The IMO Convention entered into force in 1958 and the new Organization met for the first time the following year.

1973 Agreement on International Convention for the Prevention of Pollution from Ships, including air pollution.

1978 Protocol including air pollution was added to the International Convention for the Prevention of Pollution from Ships.

1997 Annex VI added to the International Convention for the Prevention of Pollution from Ships: "Regulations ... seek to minimize airborne emissions from ... and the carbon intensity of global shipping in order to reduce its contribution to local and global air pollution and environmental problems."

2005 Annex VI of the International Convention for the Prevention of Pollution from Ships entered into force.

2008 New measurement, monitoring and certification procedures for diesel engine emissions.

2011 Mandatory measures adopted to increase energy efficiency.

2018 Initial Strategy to Reduce GHG Emissions in International Shipping was published. Its objective is to reduce carbon emissions from international shipping by 40% by 2030 and by 70% by 2050, compared with 2008. It was scheduled for review in 2023.

2022 Established enhanced fuel oil sampling and testing procedures for particulate matter.

2023 Marine Environment Protection Committee (MEPC) adopted an agreement that includes emission reduction targets compared to 2008 of 20% by 2030, 70% by 2040, and net-zero emissions "by or around, i.e. close to, 2050." The ambiguity about 2050 is one limitation. Another is that the targets are not binding.

Sources: IMO (2016, 2020, 2023a, 2023b), The Maritime Executive (2023)

Questions to Ponder

1. Which of the transportation modes is the most carbon-intensive and which the least carbon-intensive? What metric(s) do you think is (are) the most appropriate for answering the question? Why?
2. Why do you think aviation contrails have received relatively little attention—except among highly specialized experts—until recently? How would you summarize the current understanding of the problems and solutions?
3. What do you think is the best approach to addressing the emission of fossil fuel motor vehicles? What are the advantages and disadvantages compared with the alternatives?
4. For maritime shipping, what are the alternatives to the diesel fuel, which is the currently dominant fuel?
5. How do you compare the records of ICAO and IMO for addressing climate change issues? How do you explain the differences between them? Can you develop an inter-disciplinary answer?

References

Brewer, T. (2019a). Black carbon emissions and regulatory policies in transportation. *Energy Policy*, 129. https://www.sciencedirect.com/science/article/pii/S0301421519301636?via%3Dihub. Accessed May 10, 2023.

Brewer, T. (2020a). Black carbon and other air pollutants in Italian ports and coastal areas: Problems, solutions and implications for policy. *Applied Science, 10*(23), 8544. Special Issue, "Improving the Environmental Performances of Maritime Transport and Ports." https://doi.org/10.3390/app10238544. Accessed May 10, 2023.

Brewer, T. (2020b). Black carbon problems in transportation: Technological solutions and governmental policy solutions. In *Transportation air pollutants*. Springer. https://www.springer.com/us/book/9783030596903. Accessed May 10, 2023.

Brewer, T. (2020c). *Transportation air pollutants: Black carbon and other emissions*. Editor and co-author. Springer. https://www.springer.com/us/book/9783030596903. Accessed May 10, 2023.

Brewer, T. (2020d). A maritime emission control area for the Mediterranean sea? Technological solutions and policy options for a 'Med ECA'. *Euro-Mediterranean Journal for Environmental Integration*, *5*, 1. https://link.springer.com/journal/41207/volumes-and-issues/5-1. Accessed May 10, 2023.

Brewer, T. (2023). Black carbon emissions in international maritime shipping. In Y. Yamineva (Ed.), *Reducing emissions of short-lived climate pollutants: Perspectives on law, policy and science* [TBA].

Burkhardt, U., & Kärcher, B. (2011). Global radiative forcing from contrail cirrus. *Nature Climate Change*, 1. https://doi.org/10.1038/nclimate1068. Accessed July 1, 2023.

Cathcart, J., & Chen, A. (2022). Contrail mitigation: A collaborative approach in the face of uncertainty. https://rmi.org/contrail-mitigation-a-collaborative-approach-in-the-face-of-uncertainty/. Accessed May 18, 2023.

Climate Action Tracker. (2021). International Shipping, 19 July 2021. https://climateactiontracker.org/sectors/shipping/. Accessed June 2, 2023.

Climate Action Tracker. (2022). International Aviation, Sept. 22, 2022. https://climateactiontracker.org/sectors/aviation/policies-action/. Accessed June 2, 2023.

Climate Action Tracker. (2023). State of Climate Action 2022. https://climateactiontracker.org/documents/1083/2022-10-26_StateOfClimateAction2022_kR0sbBZ.pdf. Accessed June 3, 2023.

Comer, B. (2022). What if I told you cruising is worse for the climate than flying? https://theicct.org/marine-cruising-flying-may22/. Accessed July 1, 2023.

European Union Aviation Safety Agency (EASA). (2020). Report from the Commission to the European Parliament and the Council. Updated analysis of the non-CO_2 climate impacts of aviation and potential policy measures pursuant to EU Emissions Trading System Directive Article 30(4). https://eur-lex.europa.eu/resource.html?uri=cellar:7bc666c9-2d9c-11eb-b27b-01aa75ed71a1.0001.02/DOC_1&format=PDF. Accessed May 18, 2023.

Fahey, D. (2015). Aviation and climate: An update. Presentation at ICAO Headquarters. https://www.icao.int/Meetings/EGAP/Presentations/E-GAP_Session%20I_David%20Fahey.Aviation%20Climate.final.pdf. Accessed July 1, 2023.

Fleck, A. (2022). *Cruise ships are the biggest black carbon polluters*. Statistica. https://www.statista.com/chart/27353/worst-black-carbon-polluters/#:~:text=This%20reveals%20how%20disproportionately%20bad,ship%2C%20at%20only%203.5%20tonnes. Accessed July 1, 2023.

ICAO. (2016). White paper on climate change aviation impacts on climate: State of the science. https://www.icao.int/environmental-protection/Documents/EnvironmentalReports/2016/ENVReport2016_pg99-107.pdf. Accessed May 19, 2023.

ICAO. (2023a). Carbon Offsetting and Reduction Scheme for International Aviation (CORSIA). https://www.icao.int/environmental-protection/CORSIA/Pages/default.aspx. Accessed June 28, 2023.

ICAO. (2023b). Convention on International Civil Aviation—Doc 7300. https://www.icao.int/publications/Pages/doc7300.aspx. Accessed May 19, 2023.

IMO. (2016). IMO takes further action on climate change. https://www.imo.org/en/MediaCentre/PressBriefings/Pages/11-data-collection-.aspx. Accessed 28 2023.

IMO. (2023a). Brief History of IMO. https://www.imo.org/en/About/HistoryOfIMO/Pages/Default.aspx

IMO (IMO). (2023b). Index of studies related to air pollution, energy efficiency and GHG emissions from ships. https://www.imo.org/en/OurWork/Environment/Pages/IMO-Publications.aspx. Accessed June 28, 2023.

International Energy Agency (IEA). (2022). World Energy Outlook 2022. https://www.iea.org/reports/world-energyoutlook- 2022. Accessed March 31, 2023.

IPCC. (2013). Climate change: The physical science basis. Report of AR5. Accessed May 19, 2023.

IMO. (2023a). IMO climate events. https://www.imo.org/en/OurWork/Environment/Pages/IMO-climate-events.aspx

IMO. (2023b). IMO's work to cut GHG emissions from ships. https://www.imo.org/en/MediaCentre/HotTopics/Pages/Cutting-GHG-emissions.aspx

Irfan, U. (2011). World War II bomber conftrails show how aviation affects climate. *Scientific American*, 7 July 2011. http://scientificamerican.org. Accessed May 19, 2023.

Lee, D. S., et al. (2021). *Atmospheric Environment, 244*, 117834. Accessed May 19, 2023.

Liu, Z., et al. (2023). Monitoring global carbon emissions in 2022. *Nature Reviews, 4*, 205–206. https://www.nature.com/articles/s43017-023-00406-z. Accessed May 13, 2023.

Nachtigall, D., et al. (2022). The climate actions and policies measurement framework: A structured and harmonised climate policy database to monitor countries' mitigation action. In *OECD environment working papers, no. 203*. OECD Publishing, Paris. https://doi.org/10.1787/2caa60ce-en. Accessed April 17, 2023.

The Maritime Executive. (2023). Historic agreement for shipping in place. https://danishshipping.dk/en/all-press-releases/2023/historic-climate-agreement-for-shipping-in-place/. Accessed July 7, 2023.

Teoh, R., et al. (2020). Mitigating the climate forcing of aircraft contrails by small-scale diversions and technology adoption. *Environmental Science and Technology, 54*(5), 2941–2950.

Transport & Environment. (2020). Airline contrails warm the planet twice as much as CO_2, EU study finds. https://www.transportenvironment.org/discover/airline-contrails-warm-planet-twice-much-co2-eu-study-finds/. Accessed May 19, 2023.
United Nations, Sustainable Transport Conference. (2021). Fact Sheet. Climate Change. https://www.un.org/sites/un2.un.org/files/media_gstc/FACT_SHEET_Climate_Change.pdf. Accessed July 2, 2023.

Resources for Keeping Up with Developments

Climate Action Tracker (CAT). https://climateactiontracker.org/
International Panel on Climate Change (IPPC). https://www.ipcc.ch/
International Council on Clean Transportation (ICCT). https://theicct.org/
Transport & Environnent (T&E). https://www.transportenvironment.org/
International Maritime Organization (IMO). https://www.imo.org/
International Civil Aviation Organization (ICAO). https://www.icao.int/
United Nations Act Now. https://www.un.org/en/actnow/transport

Industry

8

> If we are to reach … 'net zero' emissions by the second half of this century, we are going to need solutions to both of these industries [cement and steel].
>
> Climate Action Tracker (2022)

8.1 Introduction

The industry segment consists of a combination of three energy-intensive subsectors—namely cement, steel, and petrochemicals—plus construction of buildings, roads, bridges, and other infrastructure (IPCC, 2021; Climate Action Tracker, 2022). A central feature of industry sector climate change issues is that they interact extensively with fossil fuels, electricity, transportation, and buildings. For instance, the transportation infrastructure for motor vehicles, trains, airports, and seaports consumes large amounts of cement and steel, as does the building sector.

As a result, the industry sector in total has been a large and growing source of ghg emissions. Over the two decades beginning in 2000, the industry sector was responsible for 45% of the increase in GHG emissions, and accounted for about one-third of all direct CO_2-eq emissions in 2019 (Climate Action Tracker, 2022; IPCC, 2021).

The emission patterns and trends are examined in more detail in Sect. 8.2. Cement, steel, and petrochemicals are considered, respectively, in Sects. 8.3, 8.4 and 8.5. The summary is in Sect. 8.6. Annex 8.1 describes the production processes of cement and steel.

A brief terminology note: The chapter follows IPCC use of the term "cement" to identify the industry in this segment. Technically, cement is a powder—which is used in combination with crushed rock, sand, and water to make "concrete."

T. Brewer, *Climate Change*, https://doi.org/10.1007/978-3-031-42906-4_8

8.2 Emission Patterns and Trends

Carbon dioxide emissions in the industry sector have increased from about 5 gt in 2000 to more than 9 gt in the early 2020s. In 2021 it was responsible for about 40% of total world energy consumption.

The technical summary in AR6 of the IPCC (2021) noted patterns and trends in the sector's emissions:

> Industry-sector emissions have been growing faster since 2000 than emissions in any other sector, driven by basic materials extraction and production.... GHG emissions attributed to the industrial sector originate from fuel combustions, process emissions, product use and waste, which jointly accounted for 14.1 $GtCO_2$-eq or 24% of all direct anthropogenic emissions in 2019, second behind the energy supply sector.

The following are indicators of the role of coal in the industry sector (excerpted from IEA, 2022):

- The cement and steel subsectors consume 70% of the sector's coal use.
- The sector's demand for coal doubled from 2000 to 2021.
- Coal is the largest source of CO_2 emissions in the sector; 4 Gt of CO_2 in 2021 were from coal.

In short, the industry sector is a major and growing source of ghg emissions and a significant consumer of coal. More specific information is presented in the following sections on cement, steel, and petrochemicals.

8.3 Cement

Worldwide, cement industry use of coal in 2000 and 2021 increased from 1662 to 4370 Mt of coal—an increase of 163%. Country and region data are presented in Table 8.1.

In Table 8.1 the relative size and the rate of increase in China are conspicuous. Other countries with relatively high levels in 2021 were India and several countries in Southeast Asia and the Middle East. The European Countries and Japan, on the other hand, experienced declines in their consumption of coal in the cement industry—and in the steel industry, as indicated in Table 8.2. In the US, meanwhile, use of coal in the cement industry increased only slightly over the 2000–2021 period, though it decreased in the steel industry (Table 8.2).

Technologies and Policies for Emissions Mitigation

The MIT Climate Portal (2023) has a helpful list of mitigation measures that can be undertaken, as in Box 8.1. Also see Dahmen et al. (2018); Dahmen & Muñoz (2014).

Table 8.1 Cement industry production use of coal (2000–2021)

Countries/regions	2000 (Mt)	2021 (Mt)	Percent change 2000–2021
Increased			
China	598	2365	+ 295
India	95	330	+ 247
Southeast Asia	92	265	+ 188
Middle East	69	188	+ 172
Brazil	39	65	+ 67
Russia	32	56	+ 75
South Africa	8	15	+ 88
United States	90	92	+ 2
Decreased			
European Union	222	177	− 20
Japan	81	52	− 36
World	1662	4370	+ 163

Source Excerpted by the author from IEA (2023); percentage change computed by the author

Table 8.2 Steel industry production use of coal (2000–2021)

Countries/regions	2000 (Mt)	2021 (Mt)	Percent change 2000–2021
Increased			
China	128	1024	+ 700
India	27	118	+ 337
Russia	59	76	+ 29
Southeast Asia	10	50	+ 400
Middle East	10	46	+ 360
Brazil	28	36	+ 29
Decreased			
European Union	178	153	− 14
United States	102	86	− 16
Japan	106	96	− 9
South Africa	8	5	− 38
World	849	1952	+ 130

Source Excerpted by the author from IEA (2023); percentage change computed by the author

Box 8.1: Menu of Mitigation Measures for Cement Emissions

- There are many ways to reduce or even eliminate the emissions from cement production. While some are available today, others may require more investment.
- One strategy is to use alternative fuels instead of fossil fuels to heat cement kilns. Today, these alternative fuels are mainly waste products, like used tires.
- Another strategy is to make blended cements that use less clinker. One type of blended cement, Portland-limestone cement, works as well as conventional cements but emits around 10% less CO_2.
- Low-carbon concrete can be made with blended cements and supplementary cementitious materials (SCM), which, like conventional cement, bind sand and gravel together when mixed with water. These SCMs include byproducts, such as fly ash from coal-fired power plants and granulated blast furnace slag from iron and steel production.
- Making concrete with net-zero emissions will require more ambitious actions, almost certainly including some amount of carbon capture in cement production. Fortunately, CO_2 captured when making cement (or from any other industrial process) can be "mineralized" and become part of the finished concrete itself.
- Once concrete has hardened, it also naturally absorbs CO_2 through a process called "carbonation," transforming it into a solid within the concrete. This process can offset some of the CO_2 emissions from cement production.

Source: Excerpts from MIT Climate Portal (2023).

8.4 Steel

In the steel industry, the world-wide use of coal increased by 130% from 849 Mt in 2000 to 1952 Mt in 2021 (see Table 8.2). As with cement, the largest increase was in China. By 2021, China's steel industry was consuming more than half of the world total. As with cement, India's 2021 volume was a distant second. In contrast, there were declining volumes in Europe, United States and Japan.

Technologies and Policies for Emissions Mitigation

A summary from Climate Action Tracker (2022) contrasts steel and cement with buildings and transport:

> With decarbonisation, we normally think of energy-related measures such as higher energy efficiency, electrification of demand, zero-carbon fuels and a zero-carbon electricity supply in order to move towards net-zero CO2 emissions. While such measures could set, say, the

buildings and transport sectors onto an emissions pathway compatible with the Paris Agreement's 1.5°C long term warming limit, this is not so easy for industry, especially in steel and cement production.

8.5 Petrochemicals

The petrochemical industry is different from cement and steel because it is more clearly dependent on fossil fuel feed stock and related technologies in its production processes.

Technologies and Policies for Emissions Mitigation

The petrochemical industry's production process yields thousands of products directly or indirectly. Thus, while the sources of the industry's production emissions are straightforward, identifying the end users of the product is more challenging. Beyond these generalizations, there are many tangible issues of course.

One of the issues of special interest is the manufacture, use and disposal of plastics (Gardiner, 2019). The issues include not only the ghg emissions in the production process, but also the ocean pollution that threatens fish and plant life as well as human health issues. An important feature of the measures to reduce plastic production and use, therefore is *co-benefits*. This is a reminder that analyses of the effects on climate change of any given action may have other significant direct positive effects on human health as well. The scope of cost-effectiveness and cost–benefit analysis therefore needs to be more expansive than a narrow focus only on climate change effects. Analyses of such different co-benefits is clearly a situation needing an inter-disciplinary approach that would include public health as well as climate change issues.

There is an additional complication: Plastics are also used, for instance, in vehicles to reduce their weight and thus their carbon emissions. Comprehensive calculations about the climate change consequences of reducing plastic production and use therefore require inclusion of offsetting effects. This is a reminder that analyses of the effects on climate change of any given action may include both beneficial and detrimental consequences.

As for the petrochemical industry's global patterns and trends, total world production of plastics increased from 2 million tonnes in 1950 to 460 million tonnes in 2019 (Ritchie & Roser, 2022). As for users, the transportation sector and buildings sector are only relatively small users—the former about 15% and the latter about 5% of the total. Packaging, by comparison, is about 30% of the total usage.

As for ocean pollution, the following is a useful summary (excerpted from Ritchie & Roser, 2022):

- Plastic waste generated in coastal regions is most at risk of entering the oceans; in 2010 coastal plastic waste—generated within 50 km of the coastline—amounted to 99.5 million tonnes.
- Only plastic waste which is improperly managed (mismanaged) is at significant risk of leakage to the environment; in 2010 this amounted to 31.9 million tonnes … [of which], 8 million tonnes—3% of global annual plastics waste—entered the ocean (through multiple outlets, including rivers).
- Plastics in the oceans' surface waters is several orders of magnitude lower than annual ocean plastic inputs. This discrepancy is known as the "missing plastic problem" ….. The amount of plastic in surface waters is not very well known: estimates range from 10,000s to 100,000s tonnes.

8.6 Conclusion

These three large industries are central to contemporary economies, and they pose a combination of major challenges to reduce the use of fossil fuels and address a diversity of technologies and policies that can mitigate emissions. Each industry exhibits a distinctive production process, and each offers a distinctive combination of technologies and policies that could be effective in reducing emissions. Cement is a large source of carbon emissions worldwide, but there are alternatives in existence and in progress that could yield significant reductions. Steel is much more important in commercial than residential design and construction. The central issue is whether apartment building investors and designers and office building owners will insist on new buildings that meet modern standards of energy efficiency and invest in money-saving energy efficiency projects in existing buildings. As for petrochemical plastics, the question is whether demand for plastic products will decline because of their environmental consequences.

Annex 8.1: Key Features of Cement, Steel and Petrochemical Production Processes

Few alternatives to coal are available today to produce steel, cement, and other industrial products. In cases where there may be alternative means such as the use of natural gas, they tend to be more expensive than coal in most regions. Cement plants tend to be located close to the point of use, often in cities and near big infrastructure projects, or close to limestone quarries where the main material input is extracted. … As a result, there are many more cement plants that use coal, or have the potential to use it, than steel plants. The United States is an instructive example, where cement plants are widely distributed across the country, while coal-based steel production is highly concentrated in the Midwest and Northeast regions (IPPC, 2021).

Cement

A useful summary of the production processes of cement—and concrete—follows from the MIT Climate Portal (2023; excerpts, with italics added):

Concrete is a mix of several different materials: water, fine aggregates (or sand), coarse aggregates (or gravel), chemical additives, and, most importantly, cement. *Cement* is what binds all of these ingredients together to give concrete its durability and distinctive, grey appearance.

Cement production, however, also generates most of concrete's emissions: [about 7% of all global ghg emissions]. Cement begins as crushed minerals that are heated in a kiln to make what is called "clinker." Clinker is ground into a powder, mixed with a few additives, and then blended with some other minerals to create cement.

This process creates CO_2 in two main ways. The first is the chemical reaction that occurs as clinker forms. The second is heating the kiln to temperatures above 2600 °F, *which must be done using fossil fuels.*

Steel

There are two principal paths in the steel production process—The Blast Furnace-Basic Oxygen Furnace (BF-BOF) path and the Electric Arc Furnace path (EAF).

Blast furnaces produce iron from iron ore through a "reduction process" that creates carbon dioxide. Carbon dioxide emissions are "unavoidable in this process." Electric arc furnaces use steel scrap.

In the EU, about 60% of the steel is produced using blast furnaces and about 40% using electric arc furnaces (European Steel Association (Eurofer), 2023); also see Association for Iron and Steel Technology (ASIT), 2023.

Petrochemicals

On a per capita basis, high income countries use as much as 20 times more plastic than lower income countries. Of the plastic produced globally, about 36% is for packaging, 16% for construction, and 10% for consumer goods such as toys and utensils (IEA, 2022).

Key production processes HVCs are produced either in multi- or single-product processes in the chemical sector; alternatively, they are sourced as by-products from refinery operations. Ethylene, propylene and BTX aromatics are co-produced in steam crackers. Whereas ethylene is produced almost exclusively in the chemical sector in steam crackers, propylene is sourced in large quantities as a by-product of refining operations, specifically, of fluid catalytic cracking. The majority of BTX aromatics are sourced from FCC and continuous catalytic reforming units in refineries. The dominant processes for producing propylene as a single product are propane dehydrogenation and olefin metathesis. Olefins can also be produced from methanol using the methanol-to-olefins process although this is done only in China, where abundant access to coal sufficiently lowers the cost of producing methanol. Aromatics can also be produced via a similar route, although this process is still at the demonstration phase. The key process for producing both ammonia and methanol is steam reforming of natural gas. Although the synthesis step that takes place after this process differs for each chemical—Haber Bosch synthesis for ammonia and methanol synthesis at various pressures—the need for hydrogen-containing synthesis gas is common to both. Oil feedstocks, such as naphtha, liquefied petroleum gas and fuel oil, can also be used, either with steam reforming or via a similar route to synthesis gas—partial oxidation. Again, China uniquely uses coal as a feedstock for producing both methanol and ammonia. Coal must first undergo gasification before synthesis gas can be obtained, which is highly energy-intensive.

Source: IEA (2018); also see Gardiner (2019).

Questions to Ponder

1. What are the similarities and differences in the climate change problems and solutions among cement, steel, and petro-chemicals?
2. How do climate change issues in the cement and steel industries interact with fossil fuel and transportation industries?
3. What are the "co-benefits" of reducing carbon emissions in the petrochemical/ plastic production process?
4. What are the alternatives to fossil fuels in the cement and steel industries?

References

Association for Iron and Steel Technology (AIST). (2023). Interactive steel manufacturing process. https://www.aist.org/resources/the-msts-steel-wheel/. Accessed July 3, 2023.

Climate Action Tracker. (2022). State of Climate Action 2022. https://climateactiontracker.org/documents/1083/state-of-climate-action-2022.pdf. Accessed October 9, 2023.

Dahmen, J., & Muñoz, F. (2014). Earth masonry unit: Sustainable CMU alternative. *International Journal of GEOMATE: Geotechnique, Construction Materials and Environment, 6*. https://www.researchgate.net/publication/305854401_EARTH_MASONRY_UNIT_SUSTAINABLE_CMU_ALTERNATIVE. Accessed June 30, 2023.

Dahmen, J., et al. (2018). Life cycle assessment of emergent masonry blocks. *Journal of Cleaner Production, 171*. . Accessed June 30, 2023.

European Steel Association (Eurofer). (2023). What is steel and how is steel made? https://www.eurofer.eu/about-steel/learn-about-steel/what-is-steel-and-how-is-steel-made/. Accessed July 3, 2023.

Gardiner, B. (2019). The plastics pipeline. https://e360.yale.edu/features/the-plastics-pipeline-a-surge-of-new-production-is-on-the-way. Accessed July 3, 2023.

International Energy Agency (IEA). (2018). The future of petrochemicals towards more sustainable plastics and fertilisers. https://iea.blob.core.windows.net/assets/bee4ef3a-8876-4566-98cf-7a130c013805/The_Future_of_Petrochemicals.pdf. Accessed July 3, 2023.

International Energy Agency (IEA). (2022). Coal in net zero transitions. World energy outlook special report. https://www.iea.org/reports/coal-in-net-zero-transitions. Accessed January 12, 2023.

International Energy Agency (IEA). (2023). World Energy Outlook 2022. https://www.iea.org/reports/world-energyoutlook- 2022. Accessed March 31, 2023.

IPCC. (2021). Technical summary. In *Climate change 2021: The physical science basis*. Contribution of working group I to the sixth assessment report of the intergovernmental panel on climate change. Cambridge University Press.

MIT Climate Portal. (2023). Concrete. https://climate.mit.edu/explainers/concrete. Accessed July 3, 2023.

Ritchie, H., & Roser, M. (2022). Plastic pollution. Our world in data. https://ourworldindata.org/plastic-pollution. Accessed June 30, 2023.

Resources for Keeping Up with Developments

Intergovernmental Panel on Climate Change (IPCC). https://www.ipcc.ch/
International Energy Agency (IEA). https://www.iea.org/
Our World in Data. https://ourworldindata.org/

Buildings

9

The built environment generates 40% of annual global CO_2 emissions.

Of those total emissions, building operations are responsible for 27% annually, while building and infrastructure materials and construction … are responsible for an additional 13% annually.

(Architecture 2030, 2023)

9.1 Introduction

The central climate change issues about buildings are their size, their energy efficiency, and their heating, cooling, cooking, lighting, and other technologies in the buildings; in addition, the electrical system technologies used to produce the incoming or self-generated electricity are relevant—whether fossil fuels, solar, wind or other. These issues vary significantly among countries and local areas within countries because of varying economies, government regulations, electric supply systems, building traditions and aesthetics, as well as climates. The industry sector in the preceding chapter is relevant because it includes cement and steel production, which are globally significant in building construction—and climate change emissions.

The chapter analyzes the patterns and trends in buildings' emissions in Sect. 9.2. In Sect. 9.3, the focus is on the amounts and sources of emissions, and measures to

T. Brewer, *Climate Change*, https://doi.org/10.1007/978-3-031-42906-4_9

mitigate the emissions and adapt to their effects are in Sect. 9.4. Section 9.5 summarizes the chapter. Annex 9.1 is an excerpt from a statement by an architectural group about the interdisciplinary challenges of designing more climate friendly buildings. Annex 9.2 introduces the Sufficiency, Efficiency, Renewable analytic framework applied by the IPCC.

9.2 Emissions

Emissions that can be attributed to buildings are often separated into those from construction in the materials used to build the building ("embodied" energy) and those from operating the many systems for heating, cooling, lighting, and other electrical and mechanical systems ("operational" energy). Embodied energy is directly related to the size of the building and the materials in it, such as cement, steel and plastics. The quantities of operational energy consumption and emissions depend on the carbon footprints of the technologies of the operating systems and the frequency of their use. The relative amounts of embodied and operational emissions depend on the age of the building; the older the building, the larger the operational emissions.

Embodied emissions from construction are "direct." Operational emissions are a mix of "direct" and "indirect"—for example heat from a fireplace versus electricity from afar for a window air conditioner. There is yet another feature of a building that affects its indirect contribution to climate change—namely its location (Moseman, 2023). Common examples are proximity to public transportation or convenient bicycle paths or walking paths in nearby woods. An extreme example is the difference between a house being built near the occupants' offices and/or children's schools versus a vacation home so far away that an airplane ride is needed to get there.

In general, office buildings are larger in volume and in their construction and operational emissions. New buildings often have more energy saving features than existing buildings. The buildings sector can be usefully and easily disaggregated according to a simple 2 × 2 matrix (see Fig. 9.1). There are significant differences along both dimensions in the emissions and measures to mitigate them.

Fig. 9.1 Segments of the buildings sector

Age	Function	
	Residential	Commercial
Existing		
New		

9.3 Mitigation and Adaptation Measures

The distinction between existing and new buildings is a key determinant of the cost-effectiveness of specific emission mitigation measures. As Climate Action Tracker (2023) has summarized the implications:

> Given the urgency of reducing emissions, *all new buildings* should be zero-carbon in operation (energy efficient and not reliant on fossil fuel-powered technology) while minimizing embodied emissions… Decarbonizing *existing* buildings will require a high annual rate of deep retrofits that drastically improve energy efficiency and replace equipment with zero-carbon options.
>
> Achieving zero emissions from the *existing building stock* will require leveraging building intervention points to accelerate the rate of energy upgrades (increasing energy efficiency, eliminating on-site fossil fuels, and generating and/or procuring 100% renewable energy).
>
> For full building sector decarbonization, *every existing building* will need to undergo energy upgrades involving a combination of: improvements in the energy efficiency of building operations, a shift to electric or district heating systems powered by carbon-free renewable energy sources, and the generation and/or procurement of carbon-free renewable energy.

9.3.1 Building Construction and Operational Issues

The distinction between *construction* and *operational* issues is illustrated in Boxes 9.1 and 9.2, where two significant alternative technologies are described—namely cement and heat pumps.

Box 9.1: Building Materials: Cement/Concrete

If the cement industry were a country, it would be the third largest emitter in the world (Carbon Brief, 2018).

There is a terminology issue that sometimes causes confusion in discussions of cement and concrete. Among engineers and other specialists, *cement is known to be an ingredient in concrete*, and the distinction is essential to understanding the production process. This is evident, for instance, in the following explanation from the MIT Climate Portal (2023; excerpts with italics added; also see MIT Concrete Sustainability Hub, 2023):

> *Concrete* is a mix of several different materials: water, fine aggregates (or sand), coarse aggregates (or gravel), chemical additives, and, most importantly, cement. *Cement* is what binds all of these ingredients together to give concrete its durability and distinctive, grey appearance. … Cement begins as crushed minerals that are heated in a kiln to make what is called “clinker.” Clinker is ground into a powder, mixed with a few additives, and then blended with some other minerals to create

> cement. This process creates CO_2 in two main ways. The first is the chemical reaction that occurs as clinker forms. The second is heating the kiln to temperatures above 2600 °F, *which must be done using fossil fuels.*

Nevertheless, IPCC reports and related publications generally use the term *cement as a synonym for concrete.* In this box, the sources from engineering and other experts use the term *concrete.* I have adopted that usage here to be consistent. Elsewhere, I use the IPCC's *sectoral* terminology to avoid adding to the confusion.

> Concrete is among the world's most consumed materials—second only to water. That's because its durability, affordability, and availability make it essential to countless construction projects, from bridges, to roads, to buildings MIT Climate Portal (2023).

Since concrete is used on such a large scale, it also produces large amounts of heat-trapping **greenhouse gases**, mostly from a manufacturing process that emits carbon dioxide (CO_2). Yet, the world will need concrete to build infrastructure that can cope with climate change and population growth. So the question is: how do we lower concrete's environmental impacts even as we continue to rely heavily upon it?

Potential mitigation measures (excerpts and rephrasing from MIT Climate Portal 2023):

There are many ways to reduce or even eliminate the emissions from cement production. While some are available today, others may require more investment.

Use alternative fuels instead of fossil fuels to heat cement kilns. The alternatives are mainly waste products, like used tires.

Make blended cements that use less clinker—for example portland-limestone cement, which works as well as conventional cements but emits around 10% less CO_2.

Make blended cements with supplementary cementitious materials (SCM). SCMs bind sand and gravel together when mixed with water; they include byproducts, such as fly ash from coal-fired power plants and granulated blast furnace slag from iron and steel production.

Include some amount of carbon capture in cement production. The captured CO_2 can be "mineralized" and become part of the finished concrete. This does not include pumping any of the captured carbon underground.

Hardened concrete naturally absorbs CO_2 through a process called "carbonation," which transforms it into a solid within the concrete. This process can offset some of the CO_2 emissions from cement production.

For further information about emissions and alternatives to traditional concrete, see the concept for "carbon-negative concrete" MIT Climate Portal (2023), Carbon Brief (2018), Dahmen and Muñoz (2014), Dahmen et al. (2018), Mehta and Meryman (2009), Muñoz et al. (2015).

9.3.2 Operational Issues: Space Heating and Cooling

Space heating and cooling are significant sources of emissions, and they are a direct function of the amount of space. See Box 9.2 for a description of how the use of heat pumps for space heating *and cooling* can reduce emissions.

Box 9.2: Heat Pumps

Heat pumps are widely recognized as a key clean energy technology in the energy transition. While the global heat pump market has expanded significantly, more than doubling in some countries in a single year, expanded policy support will be needed to build confidence in the technology and meet climate goals.

Heat pumps are a low-carbon heating technology with the potential to deliver large-scale reductions in carbon emissions from building heat. They use electricity to move heat from ambient outside air, water, or ground to a building's interior and to heat water. This process is highly efficient, with heat pumps delivering three to five units of heat for each unit of electricity needed to run them. In addition to being highly efficient, heat pumps also use predominantly renewable thermal heat rather than relying on combusting fossil fuels. As much of the useful heat from a heat pump comes from inexhaustible environmental sources, 70–80% of energy provided by an average heat pump is renewable. Even with today's electricity mix, which tends to include fossil fuels such as coal, heat pumps can reduce emissions in most of the world's regions, which together made up 96% of global heating energy demand in 2015. When the electricity used to drive the electric compressor is produced from low-carbon sources, nearly all of the useful heat provided becomes low or even zero carbon.

Heat pumps have been repeatedly identified as a key, cost-effective solution for tackling the carbon emissions associated with keeping buildings warm at international, regional and national levels. The costs of low carbon electricity have also declined significantly over the last decade, bolstering the case for electric heat pumps.

Europe has been leading the shift, with double digit growth rates from 2015 onwards. Other countries, including China, are seeing comparable developments more recently. Similar to the market for electric vehicles, the increasing recognition of this technology in energy strategies and the growing use of clean heating are expected to lead to much faster and potentially much more significant deployment and use of heat pumps around the world.

Source: Rosenow et al. (2022), also see Climate Action Tracker (2023).

Box 9.3: Government regulations: building performance standards

As … codes push new buildings to meet zero-energy and zero-carbon targets in their operations, corresponding efforts to reduce embodied energy and carbon from building construction materials must be pursued to achieve building-sector decarbonization goals. Moreover, at the intersection of the buildings and industrial sectors, initiatives to decrease embodied carbon in building materials will increase demand for low-carbon products, thereby improving the business case for manufacturing industries [including concrete, steel, and plastic]. …

One challenge of incorporating embodied carbon provisions into model building codes lies in the complexity of the codes-development process, which requires standardized methodologies at the national and international levels, robust technoeconomic analyses and modeling, and thorough evaluations of the codes' impact on various stakeholder groups (including manufacturers, building owners and developers, builders, and occupants). …

[T]he multi-layered solutions that will ensure the success of policies … across complicated supply chains must be rigorously developed and tested; they must also be applicable to a dizzying array of products. Leveraging our decades of experience in working with and convening industry, [the American Council for an Energy-Efficient Economy] is collaborating with leading manufacturers and their trade associations to build business cases for reducing embodied carbon in building materials. These efforts will also increase the availability of low-embodied-carbon products and materials in related non-building markets such as transportation and infrastructure.

Source: American Council for an Energy-Efficient Economy (ACEEE) (2023).

9.4 Conclusion

The buildings sector is highly dependent on construction materials produced by the cement, steel, and petrochemical industries featured in the prior chapter. Since all these building materials are manufactured by carbon intensive production processes, the adoption of alternative building materials is essential for the buildings sector to reduce its emissions. Climate friendly operational systems for heating and cooling buildings and for consuming electricity are already available—and at increasingly competitive prices that are making them attractive in many countries. Yet, despite the mix of economic incentives and regulatory requirements of building codes and sustainable energy subsidies, the speed and scope of the changes in the building sector to date have not been sufficient to meet widely accepted emission reduction targets.

Annex 9.1: Building Design: Interdisciplinary Challenges for Architects

Climate change affects every person, every project, and every client. The impacts are all-inclusive, with no respect for borders or boundaries—and are felt first and hardest by our most vulnerable populations. Rising sea levels, extreme weather events, and the degradation of natural resources are a direct result of increased carbon levels.... Because more than 40% of U.S. greenhouse gases can be attributed to the building industry—during construction, embodied in concrete, metals, and polymers, and through everyday processes such as heating, cooling, and lighting—architects have the ability to lead the change our planet needs.

This represents a dramatic and unprecedented change in [The American Institute of Architects'] focus, unequaled in its 162-year history, and comes with a unique set of challenges. We need to reach a diverse audience of architects, partners, and clients. ...

We cannot address these challenges alone. We must learn from others' research and integrate diverse expertise into our collective growth. Public health professionals, insurers, supply chain analysts, and so many others have expertise that complements that of architects, and their work informs design solutions. ...

The AIA's is focusing efforts on three overarching goals to address the sources of climate change, adapt to the impacts of climate change, and to lead and transform the architectural industry:

1. Mitigating the sources. Establish the relevance and importance of the building sector and architectural practice in climate mitigation solutions. By actively addressing the building industry's footprint as a primary contributing source of operational and embodied carbon, architects must play a major role in catalyzing the industry by advancing zero-carbon projects, products, policies, initiatives, research, and education. ...
2. Adapting to the impacts. Design buildings and communities to anticipate and adapt to the evolving challenge of climate change. By addressing the impacts of climate change in every design solution, our spaces, buildings, structures, and communities become more functional and high performing. ...

3. Catalyzing architects to act. Lead meaningful change and contribute to climate solutions in partnership with our global community. The challenges around embodied carbon and existing buildings, new building design, renewable energy, and electrification go hand-in-hand with opportunities.

Source: Architecture 2030 (2023)

Annex 9.2: Sufficiency, Efficiency, Renewable Analytic Framework

Use of the Sufficiency, Efficiency, Renewable (SER) framework aims to reduce the cost of constructing and using buildings without reducing occupants' well-being and comfort.

Sufficiency measures tackle the causes of GHG emissions by limiting the demand for energy and materials over the lifecycle of buildings and appliances. Sufficiency policies are a set of measures and daily practices that avoid demand for energy, materials, land and water.

Sufficiency interventions do not consume energy during the use phase of buildings and do not require maintenance or replacement over the lifetime of buildings. Sufficiency interventions that have been implemented in leading municipalities include:

> Density, compacity, bioclimatic design to optimise the use of nature-based solutions, multi-functionality of space through shared space and to allow for adjusting the size of buildings to the evolving needs of households,
>
> Circular use of materials and repurposing unused existing buildings to avoid using virgin materials,
>
> Optimisation of the use of buildings through lifestyle changes,
>
> Use of the thermal mass of buildings to reduce thermal needs,
>
> Moving from ownership to usership of appliances.

Sufficiency differs from efficiency. Sufficiency is about long-term actions driven by non-technological solutions, which consume less energy in absolute terms. *Efficiency* is about continuous short-term marginal technological improvements.

Source: Rearranged excerpts and rephrasing from IPCC (2022, Chapter 9).

Questions to Ponder

1. What are the major contributing factors to a building's carbon footprint?
2. How do *heat* pumps reduce energy consumption for *air conditioning* systems?
3. What is source of the electricity in the building you are in now or another one that you are in often?
4. What are some transportation/urban planning issues related to building issues?

References

American Council for an Energy-Efficient Economy (ACEEE). (2023). *Embodied carbon in buildings and infrastructure.* https://www.aceee.org/embodied-carbon-buildings-and-infrastructure. Accessed July 4, 2023.

Architecture 2030. (2023). *The built environment.* https://architecture2030.org/why-the-building-sector/. Accessed July 4, 2030.

Climate Action Tracker. (2023). *Buildings.* https://climateactiontracker.org/. Accessed January 12, 2023.

Carbon Brief. (2018). *Why cement emissions matter for climate change.* https://www.carbonbrief.org/qa-why-cement-emissions-matter-for-climate-change/. Accessed July 4, 2030.

Dahmen, J., & Muñoz, F. (2014). Earth masonry unit: sustainable CMU alternative. *International Journal of GEOMATE: Geotechnique, Construction Materials and Environment, 6.* https://www.researchgate.net/publication/305854401_EARTH_MASONRY_UNIT_SUSTAINABLE_CMU_ALTERNATIVE. Accessed January 12, 2023.

Dahmen, J., Kim, J., & Ouellet-Plamondon, C. M. (2018). Life cycle assessment of emergent masonry blocks. *Journal of Cleaner Production, 171.* https://www.sciencedirect.com/science/article/abs/pii/S0959652617323351. Accessed July 4, 2030.

IPCC. (2022). *Climate change 2022. Mitigation of climate change. Technical summary.* Working Group III contribution to the Sixth Assessment Report of the Intergovernmental Panel on Climate Change. https://www.ipcc.ch/report/ar6/wg3/downloads/report/IPCC_AR6_WGIII_TechnicalSummary.pdf. Accessed June 12, 2023.

Mehta, P. K., & Meryman, H. (2009). Tools for reducing carbon emissions due to cement consumption. *Structural Magazine, 16.* https://www.structuremag.org/wp-content/uploads/2014/08/C-BB-SustainableConcrete_MehtaMeryman-Jan091.pdf. Accessed July 4, 2030.

MIT Climate Portal. (2023). *Concrete.* https://climate.mit.edu/explainers/concrete. Accessed July 3, 2023.

MIT Concrete Sustainability Hub. (2023). https://cshub.mit.edu/. Accessed June 13, 2023.

Moseman, A. (2023). MIT climate portal. How much CO_2 is emitted by building a new house? https://climate.mit.edu/ask-mit/how-much-co2-emitted-building-new-house. Accessed July 3, 2023.

Muñoz, J. F., Easton, T., & Dahmen, J. (2015). Using alkali-activated natural aluminosilicate minerals to produce compressed masonry construction materials. *Construction and Building Materials, 95.* https://www.sciencedirect.com/science/article/abs/pii/S0950061815301665. Accessed July 4, 2023.

Rosenow, J., Gibb, D., Nowak, T., & Lowes, R. (2022). Heating up the global heat pump market. *Nature Energy, 7.* https://doi.org/10.1038/s41560-022-01104-8. Accessed January 12, 2023.

Keeping up with Future Developments

Architecture 2030. https://architecture2030.org/
Carbon Action Tracker. https://climateactiontracker.org/
Intergovernmental Panel on Climate Change (IPCC). https://www.ipcc.ch/
International Energy Agency (IEA). https://www.iea.org/
IRENA. https://www.irena.org/Energy-Transition/Country-engagement/Climate-Action

10 Agriculture, Aquaculture, Food, Forests and Other Land and Ocean Uses

10.1 Introduction

This sector is the most heterogeneous of the standardized IPCC sectors, though there is an underlying similarity among the subsectors—namely, they involve land use and/or ocean use. Beyond this feature of the sector, it is unique because of its importance as a *sink* for carbon emissions. In fact, many of the most important issues about the sector concern the capacity of the forests and oceans to continue to absorb enormous quantities of carbon each year. It should also be noted that this chapter includes aquaculture and other ocean-related issues that are not always included in "AFOLU" topics—i.e. "Agriculture, Forestry and Other Land Use" issues. However, there are underlying similarities in the ocean issues and the standard AFLOU issues.

In any case, the AFLOU sector was responsible for 13–21% of total global anthropogenic GHG emissions during the decade 2010–2019 (IPCC, 2021b).

Of course, all of the sub-sectors are exposed to multiple effects of climate change, and these are discussed in the individual subsectors. Section 10.2 focuses on agriculture, Sect. 10.3 on forests, Sect. 10.4 considers aquaculture and the role of oceans as carbon sinks. Section 10.5 is a summary of the chapter. (For information about "other land use" issues, see UN (2023).)

10.2 Agriculture

10.2.1 Agricultural Vulnerabilities to Climate Change

Agriculture is exposed to many of the direct effects of climate change, including droughts, floods, and variable precipitation rates. Developing countries are particularly vulnerable because of the importance of agriculture in their societies, their relatively low income levels, and their exposure to droughts (Cline, 2007;

T. Brewer, *Climate Change*, https://doi.org/10.1007/978-3-031-42906-4_10

FAO, n.d.; IPCC, 2021; Rosa & Gabrielli, 2023). Population growth also entails increasing demand for agricultural production and thus increasing vulnerabilities to climate change.

10.2.2 Agricultural Adaptation Measures

A Food and Agriculture Organization (FAO) program has been developed to take action "before a crisis escalates into an emergency [so that] disaster losses and emergency response costs can be drastically reduced. [In addition], early action also strengthens the coping capacities of at-risk populations" (FAO, n.d.). Fourteen countries in Africa, the South Pacific, Asia and Central America are … specifically targeted due their increased risk to extreme weather and a subsequent negative effect on vulnerable people. Another 19 countries are classed as facing moderate risk. Details about some countries covered by the program are in Box 10.1.

Box 10.1: Examples of the FAO Early Warning—Early Action System

In Somalia, riverbanks are being reinforced and sandbagged.

In Malawi, the [FAO] is assisting governments in the preparation of a food insecurity response plan.

In Zimbabwe, the FAO is … responding to the foot and mouth disease outbreak where 5.4 million doses of vaccines are still required, and it has also prepared a drought mitigation program.

In Central America, [a program has been developed in order] to increase the resilience of households, communities and institutions to prevent and address disaster risks that affect agriculture and food and nutrition security.

Source: Excerpted by the author from FAO (2015); also see FAO (n.d.)

10.2.3 Agricultural Emissions

The global food system contributes about one-third of annual global CO_2 equivalent ghg emissions—for instance, 30% in 2020 (Rosa & Gabrielli, 2023). About one-fifth of the food system share, however, is attributed to transportation and other processes not included in this chapter's sector. Within the total food system, agriculture contributes about 40%.

An important feature of agriculture's emissions is that they are dominated by methane and nitrous oxide, which are both highly potent. The proportions of agriculture's total are that methane contributes 54%, nitrous oxide 28% and carbon dioxide 18% (Rosa & Gabrielli, 2023).

Table 10.1 Atmospheric lifetimes and GWPs of agricultural emissions

Emissions	Average atmospheric lifetime (years)	GWP 20 years/100years
Methane (CH_4)	11.8	81/27
Nitrous oxide (N_2O)	109	273/273
Carbon dioxide (CO_2)	100	1/1

Source IPCC (2021a)

Table 10.2 Sources of agriculture emissions

Sources of emissions	Global total: Gt CO_2-eq (2020)	Emission details
Enteric fermentation	2.85	Methane from digestive system of cattle
Manure management	0.28	Methane from decomposition
On-farm energy use	1.03	Carbon dioxide from electricity and machinery fuel
Synthetic fertilizers	1.01	Nitrous oxide and carbon dioxide
Rice cultivation	0.69	Methane from flooded rice fields
Crop residues	0.23	Methane from decomposition and burning

Source Excerpted from Rosa and Gabrielli (2023)

As for the future, a report of the IPCC (2022) concluded that:

> Taken together, recent research shows that achieving global food security in the coming decades, while limiting warming to 1.5°C, cannot be done without significant changes to food production and consumption. Shifting demand, increasing productivity, and changing on-farm practices and technologies, combined, are necessary to reduce global emissions ….

Table 10.1 indicates the 20-year and 100-year global warming potentials of the three main carbon emissions in agriculture. The 20-year GWP for methane is especially notable for agricultural emissions and for global efforts to reach net zero.

Table 10.2 summarizes the sources of the emissions.

10.2.4 Measures to Mitigate Agricultural Emissions

There are diverse kinds of emission mitigation measures, as summarized in Box 10.2.

Box 10.2: Mitigation Measures

Market-based instruments, which either make producers pay for emissions, through a carbon tax or emissions trading scheme or pay for abatement, through subsidies or the provision of offset credits in compliance and voluntary carbon markets.

Regulations that prohibit certain activities, impose standards or specify technologies or products to use in production. Because of the high degree of heterogeneity in the agriculture sector, direct regulation through uniform standards is unlikely to be a cost effective option.

Provision of affordable lines of credit to leverage abatement through investments rather than payments.

Information provision, including via government advisory services.

Government certification schemes to provide access government subsidies based on standardized quantification methodologies.

Source: Excerpted and re-phrased from OECD (2022)

Box 10.3: Agriculture Project in Kenya that Decreases Carbon Emissions and Increases Productivity

An international partnership based in Kenya that includes participants from Belgium, Germany and Brazil has created a cooperative agricultural project that increases the productivity of the land while reducing carbon emissions. The project combines locally available soil organic matter, biochar and enhanced weathering to rapidly restore healthy fertile topsoil for growing more food with less synthetic fertilizer and irrigation. Furthermore, in the process, these nature-based solutions absorb large amounts of carbon from the atmosphere for many years.

Soil organic matter (decomposing plant or animal tissue) and biochar (from feedstocks such as tea prunings and bagasse) contribute to soil productivity, increase its capacity to store water, and reduce denitrification.

When rocks on the surface of the soil are broken down or dissolved, for example, by acid rain or change in temperature, they re-mineralize the soil. Accelerating that is known as "enhanced weathering." Kenya's abundant shallow basalt deposits from volcanic activity, and its quarry waste as the country expands its roads and built environments, can be used for this purpose. Soil re-mineralizers are already producing in Brazil, where the rocks come from metallurgical mining.

The supply of crushed rocks and biochar can be expensive, and farmers transitioning to regenerative practices require financial support; however, emissions offset certificates could reduce costs. Farmers can also benefit from recurrent savings on input use, and yield increments. In addition, the food for humans or livestock will be more nutritious.

Source: Personal communication from Aarti Shaw to the author, with editing by the author.

10.2.5 Agricultural Carbon Sinks

The potential of agricultural carbon sinks is currently being *reduced* by many common farming operations noted in Cho (2018): "Agricultural practices that disturb the soil—such as tilling, planting mono-crops, removing crop residue, excessive use of fertilizers and pesticides and over-grazing—expose the carbon in the soil to oxygen, allowing it to burn off into the atmosphere."

Yet, there are numerous ways that carbon sequestration measures *could* enhance the use of agricultural land as carbon sinks (OECD, 2022):

Mineral carbonation of soil
Erosion control
Fire management
Grazing land management
Improved rotations
Perennial crops
Management of organic soils
Nutrient management
Organic resource management
pH management
Tillage management
Water management.

The OECD (2022) estimates that such measures have the potential to offset 2–4% of total global GHG emissions by the end of the century.

10.3 Forests

There is a useful, balanced statement about the status and future of the world's forests in the UN Climate Change News (2023):

> Forests globally are under immense pressure. The Intergovernmental Panel on Climate Change's 6th Assessment Report … finds that forests are under severe threat, while sustainably managed forests play an important role in both reducing greenhouse gas emissions … and adapting to the impacts of climate change.

> Forests also provide many other ecosystem goods and services, such as protecting biodiversity and maintaining water supply and soil quality, as well as contributing to the sustainable livelihoods of millions of people around the world.
>
> The risk of wildfires is also increasing, and … governments are urged … to rethink their approach to extreme wildfires.

These are therefore among the key questions about the forests: How much are the forests' capacities to serve as carbon sinks being threatened? How and by whom? What policies and practices can protect them—and even augment their carbon sink capacities?

Terminology

There is a terminological distinction that is essential to understanding the past, current, and future of the relationship between the world's forests and climate change. The distinction concerns the terms "reforestation" and "forest restoration" (Climate Action Tracker, 2022; IPCC, 2022). "Reforestation" refers to a process whereby forests grow up in non-forest land where there have been forests in the past. The reforestation can be the result of natural processes or human intervention. "Forest restoration" is a more extensive process than increasing the number of trees; its purpose is re-establishing the forests' "ecological functions" (IPCC, 2022). Another distinction is between "natural forests" and "plantation forests." These conceptual distinctions are not only important to understanding the processes, they are also sometimes a source of uncertainty and even confusion about the implications of the available data.

Emissions and Carbon Sinks

During the year 2022, the total global "tropical primary forests" declined by 10% in "natural forests" (Global Forest Watch/World Resources Institute, 2023; Hodgson et al. 2023). Forest fires are an important cause of the decline in many countries. Over the period from 2001 to 2022, about 27% of the global loss was from fires. In 2022, fires were responsible for about 23% of the losses. The amount being destroyed by fires in 2022 was twice as much as 20 years previously (Global Forest Watch/World Resources Institute, 2023). An extreme case was Ghana's forest loss in 2022 that was 70% more than 2021. At the same time, in other countries such as Indonesia and Malaysia, there have been declines in the rate of loss from fires.

In some countries, criminal activities have also been significant contributors (Hodgson et al. 2023).

As for the future, a large-scale study (United Nations Environment Programme, 2022) concluded that:

> By the end of the century, the likelihood of catastrophic wildfires events will increase by a factor of 1.31 to 1.57. Even under the lowest emissions scenario, we will likely see a significant increase in wildfire events.

The world's forests not only pose important issues about annual changes in their sizes and the causes of the changes; they also pose important issues over the long term for their capacities to serve as sinks. Forests acting as carbon sinks are estimated to store about 45% of *terrestial* sinks (Waring et al. 2020). These land-based sinks are thus separate from the ocean-based sinks discussed in the chapter section below.

Policies and Practices to Protect the Forests

A series of international agreements have established frameworks for governments and other interested organizations to take actions to protect the forests. The international agreement on Reducing Emissions from Deforestation and forest Degradation (REDD+) refers to a list of activities which countries may use to slow, halt and reverse forest cover and carbon loss. It was initially agreed at COP21 in Paris in 2015, which was then supplemented with a series of agreements through COP26 in Glasgow in 2021. At COP26, there was an agreement in which world leaders endorsed the Glasgow Leaders Declaration on Forests and Land Use; they agreed to halt and reverse forest loss (and land degradation) by 2030.

Implementations with tangible actions were evolving a year later. In particular, pledges made by the end of 2022 of USD 12 billion during 2021 and 2025 were "still only a fraction of what is needed to meet 2030 goals," according to the Forest Declaration Platform (2023), which is sponsored by the UN Development Programme and several NGOs. Whether the program would become more sizable was an open question as of mid-2023.

10.4 Aquaculture

Aquaculture can be defined simply as farming in water, including breeding, raising, and harvesting fish, shellfish, and aquatic plants. It occurs in oceans and estuaries as "marine" aquaculture and in "freshwater" locations. This discussion focuses on oceans because oceans cover 70% of the earth's surface and because the relative actual and potential sizes of oceanic aquaculture are large and include many species. The emphasis here is how climate change affects aquaculture.

10.4.1 El Nino and La Nina

Periodic occurrences of El Nino and La Nina, which are connected to changing sea surface temperatures, can have major effects on agricultural production. El Ninos contribute to the frequency and intensity of extreme warm weather events and thus pose crop failure and food security problems (Ahuja, 2023; WMO, 2023; US NOAA, 2023a, 2023b).

El Nino and La Nina events occur as of the El Niño/La Niña Southern Oscillation (ENSO) in the Pacific Ocean, which is otherwise in an ENSO-neutral stage. There is much variability and hence uncertainty in the timing of the events, with

El Nino events happening between two and seven years. A single event can last as long as a year and a half.

Although the ENSO conditions pertain to the Pacific Ocean, there are widespread consequences that include changes in the Arctic jet stream, for instance, and hence weather patterns in many parts of the world. The temporal and geographic variations in the US summarized in Box 10.3 illustrate the variations and the regional effects far from the Pacific Ocean, as well as those on the Pacific coast.

Box 10.4: Temporal and Geographic Variations of El Nino and La Nina Effects
El Nino's influence on the U.S. is weak during the summer and more pronounced starting in the late fall through spring. By winter, there is an 84% chance of greater than a moderate strength El Nino, and a 56% chance of a strong El Nino developing. Typically, moderate to strong El Nino conditions during the fall and winter result in wetter-than-average conditions from southern California to along the Gulf Coast and drier-than-average conditions in the Pacific Northwest and Ohio Valley. El Nino winters also bring better chances for warmer-than-average temperatures across the northern tier of the country.

A single El Nino event will not result in all of these impacts, but El Nino increases the odds of them occurring.

Source: US NOAA (2023a, 2023b)

More generally, the Food and Agriculture Organization (FAO) has summarized the consequences of an El Nino on agriculture and aquaculture, as in Box 10.3.

Box 10.5: FAO Summary of El Nino Effects on Agriculture and Aquaculture

Agriculture is one of the main sectors of [an] economy that [can] be severely affected by the El Niño phenomenon. While drought is the main threat to food production, El Niño can also cause heavy rains, flooding or extremely hot or cold weather. This can lead to animal disease outbreaks, including zoonosis and food-borne diseases, as well as plant pests and forest fires. In [some] El Niño events, people whose livelihoods depend on fisheries have been heavily affected in certain areas.

Source: Excepted by the author from FAO (2015); also see FAO (n.d.)

10.4.2 Aquaculture Vulnerabilities to Climate Change

Three dimensions of climate change affect aquaculture: increasing water temperature, increasing acidification, and changing currents. Acidification refers to the process whereby the oceans absorb carbon dioxide, which lowers the pH of the water and thus poses a danger to coral reefs. Changes in currents are more complex and regionalized, but with global consequences; a well-known example is the Atlantic Meridional Overturning Circulation (AMOC) system (see Chap. 2 for more information about the AMOC).

A key question about the effects of these dimensions of climate change is thus: How are the various species of fish affected? In many cases, the stocks of fish are being reduced. Shellfish, for example, are affected by having their shells underdeveloped by acidification. Annex 10.1 presents a case that combines the effects of increasing water temperature, acidification and changing currents on lobsters and oysters in the US northeast coast of the Atlantic Ocean.

10.4.3 Oceans as Carbon Sinks

The oceans have been absorbing about one-fourth of global carbon dioxide emissions since the beginning of the industrial revolution, but their capacity to continue this process is declining (Woods Hole Oceanographic Institution, 2023). The problem is that the algae and bacteria, which are forms of phytoplankton and absorb the carbon that falls into the oceans, are being threatened by plastics and other pollutants. Thus, while the upper limit of the oceans' capacity to serve as a sink is being reduced by the increasing volumes of emissions landing in the water, the volume of the phytoplankton that can absorb the carbon is being reduced. Any decline in the oceans' role as carbon sink places more pressure on the forests to compensate by absorbing more.

10.5 Conclusion

The land-based subsectors of this chapter are affected directly by climate change, especially by droughts and extreme heat episodes. The ocean-based subsectors are affected by warming water, acidification, and changing currents. Food security is at risk in the agriculture and aquaculture subsectors, where there are catastrophic health and economic consequences for millions of people as producers and consumers. At the same time, the forests and oceans serve as significant carbon sinks that reduce the carbon dioxide and other carbon pollutants in the atmosphere. The destruction of the forests and the declining available carbon budget of the oceans are both increasing climate related risks. There are many available mitigation and adaptation measures, but they are not being implemented with sufficient speed or scale to meet the challenges of the sector.

Annex 10.1: The Case of Changes in the Gulf of Maine

The Gulf of Maine includes the North Atlantic coastal area from Massachusetts in the US to Nova Scotia in Canada. The Gulf is among the fastest warming major bodies of water in the world's oceans, and it has also experienced significant changes in its currents and its salinity.

It is important to understand the wider regional and global causes and effects, including how and why the changes are occurring. The case is an instructive example of how changes in oceans that are induced by climate change can have wide-ranging causes and effects in many parts of the world. (Sects. 10.4.2 and 10.4.3 of this chapter have additional information about changes in ocean temperatures, salinities and currents in general. (Annex 2.2 in Chap. 2 has additional information about the causes and consequences of global warming in the Arctic.)

Global Context

Average Arctic temperatures have increased by about four times the global mean temperature in recent years. The melting glaciers on Greenland discharge enormous volumes of fresh water into the Arctic Sea; that fresh water, in turn, changes Atlantic Ocean salinity, temperature and currents. Greenland's glacial melt also contributes significantly to global sea-level rise (Voosen, 2020). Black carbon and methane, as well as carbon dioxide, are all major global warming agents in the Arctic. The emissions come from as far away as China, from international cargo and cruise ships that transit the Arctic Ocean during the summer when the sea ice melts, and from local sources year-round as well.

There are also global consequences of the warming in the Arctic region. The consequences include sea level rise far beyond the Arctic Circle, changes in the Atlantic Gulf stream that in turn affect European weather, changes in weather patterns in much of the northern hemisphere resulting from changes in the Arctic "jet stream." There is also the potential of a "tipping point" that could occur when thawing of the Arctic tundra would cause release of underground arctic methane, a tipping point that would have world-wide consequences (Arctic Council, 2022; Intergovernmental Panel on Climate Change, 2019).

Local and Regional Effects

Local effects, of course, do not arise at such a large scale. However, they are tangible illustrations of how *global* climate change issues are already posing issues for local governments, businesses, and residents. Local governments, for instance, have begun to develop projects to protect transportation infrastructure and buildings from sea level rise. The emphasis here for this chapter, though, is the effects on aquaculture.

Communities on the state of Maine coast are exposed to the climate change effects on the Gulf of Maine, which is a major source of seafood, including lobsters in particular. Some lobsters are harvested in aquaculture facilities and others in individual commercial fishing boats. In addition, a series of commercially successful oyster farms have been established in the Damariscotta River, which is one of many rivers with a Gulf of Maine estuary. Maine is one of a few major sources of oysters in the US, and there are more than a dozen oyster farms in the Damariscotta River in particular. They currently produce about 80 percent of the state's oyster harvest.

This annex case study, therefore, focusses on the effects of climate change on these seafood sources. The answers are based on a combination of standard governmental data sources, published research, and the author's exchanges with regional experts who have specialized knowledge of many aspects of the issues.

Temperature Changes

A study by Balch et al. (2022) concluded that:

A new synthesis of two decades of data has elucidated the startling transformation of the warming Gulf of Maine. Gathered and analyzed by researchers at Bigelow Laboratory for Ocean Sciences, the study shows many trends that all point to an overarching pattern. The Gulf of Maine is being increasingly influenced by warm water from the North Atlantic, and this is changing the very foundation of its food web (Balch et al. 2022; also see American Geophysical Union, 2022; and Bricknell, 2013).

Figure 10.1 compares the temperatures in the Gulf of Maine with the world global average over the four-decade period from 1980 to 2020. Over that period, the *decadal* increase for the Gulf of Maine was 3.2 *times* greater than the global oceanic average. Source: There are important year to year variations in the temperature time and differences across regions, but the basic trend and its causes are apparent. There have also been significant changes in acidity and salinity.

Acidity Changes

Because the development of lobster shells and oyster shells is affected by the acidity of the surrounding water, the increasing acidity of the Gulf of Maine is also an issue for the future of those two important resources in the local economy. As with temperature, there are ranges of acidity that are optimal for shell formation.

Salinity Changes

The loss over two decades of up to 65% of its productivity from phytoplankton is a significant element of the changing Gulf. Because phytoplankton are at the bottom of the marine food web, this trend is consequential for fishing and aquaculture.

Implications

There is no doubt that conditions in the Gulf of Maine and its estuaries have been changing and that the changes are pertinent to the prospects for the fishing industry and thus the societies that depend on them. It is not yet clear, however, what the mix of potentially beneficial and detrimental consequences will be.

Sources: Balch et al. (2004, 2008, 2022), Balch (2016), Bricknell (2013), Gulf of Maine Research Institute (2023a, 2023b) I am especially indebted to the following for their expert contributions to the Annex: Barney Balch, Bigelow Laboratory

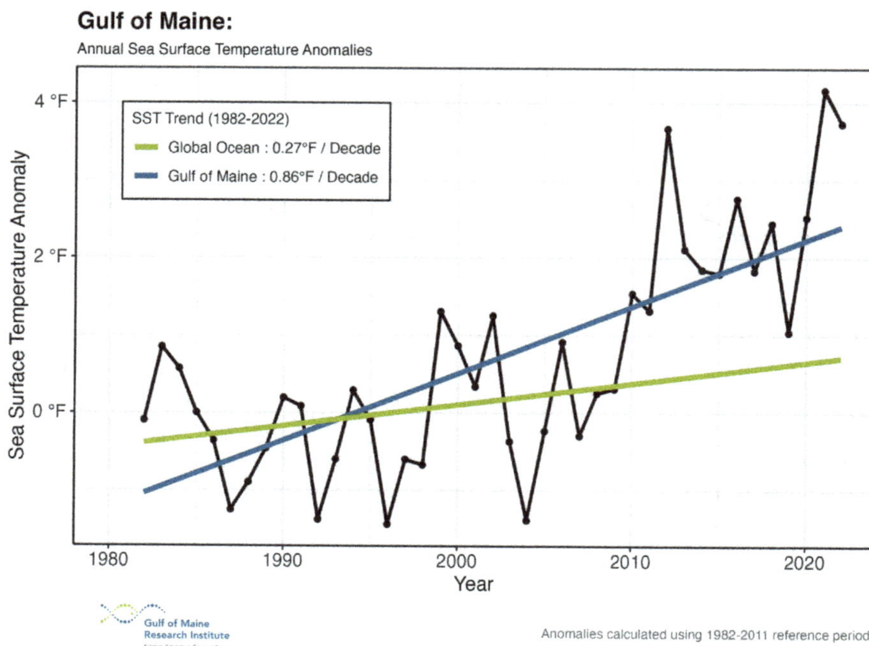

Fig. 10.1 *Source* Gulf of Maine Research Institute (2023a; also see 2023b); used with permission

for Ocean Sciences; Damian Brady, University of Maine, School of Marine Sciences; Jennifer Brewer, University of New Hampshire, Geography Department; Sara Gladu, Coastal Rivers Conservation Trust.

Questions to Ponder

1. What is the earth's carbon budget? Why is it important? What role do oceans have in it?
2. Are farms net carbon sources or sinks?
3. Why do forests in Brazil get so much attention from climate scientists?
4. What are the changes in the world's oceans that affect aquaculture? Are the changes beneficial or detrimental?

References

Ahuja, A. (2023). El Niño's arrival spells out the stark choices on climate. *Financial Times*, 14 June. https://www.ft.com/content/08cfda0c-62ca-4f4d-8677-2fa747f1a2bd. Accessed June 14, 2023.

American Geophysical Union (2022). https://news.agu.org/press-release/changes-in-gulf-of-maine-waters-impact-food-web-from-bottom-to-top/. Accessed 27 June 2023.

Arctic Council (2022). Black Carbon and Methane. Accessed at https://www.arctic-council.org/about/task-expert/egbcm/ on 18 November 2022.

Balch, B., et al. (2004). A multi-year record of hydrographic and bio-optical properties in the Gulf of Maine: I. Spatial and temporal variability. *Progress in Oceanography,* 63 (2004) 57–98. https://www.sciencedirect.com/science/article/abs/pii/S0079661104001004. Accessed 27 June 2023.

Balch, B., et al. (2008). Space–time variability of carbon standing stocks and fixation rates in the Gulf of Maine, along the GNATS transect between Portland, ME, USA, and Yarmouth, Nova Scotia, Canada. *Journal of Plankton Research*, 30. https://agupubs.onlinelibrary.wiley.com/doi/full/10.1002/2015gb005332. Accessed 27 June 2023.

Balch, B., et al. (2022). Changing Hydrographic, Biogeochemical, and Acidification Properties in the Gulf of Maine as Measured by the Gulf of Maine North Atlantic Time Series, GNATS, Between 1998 and 2018. *Journal of Geophysical Research: Biogeosciences*, 127. https://doi.org/10.1029/2022JG006790. Accessed 27 June 2023.

Balch, W. (2016). Toward a quantitative and empirical dissolved organic carbon budget for the Gulf of Maine, a semi enclosed shelf sea. *Global Biogeochemistry Cycles*, 30. doi: https://doi.org/10.1002/2015GB005332. Accessed 27 2023.

Balch, W. B., et al. (2012). Step-changes in the physical, chemical and biological characteristics of the Gulf of Maine, as documented by the GNATS time series. *Marine Ecology Progress Series.* https://doi.org/10.3354/meps09555. Accessed 27 June 2023.

Bricknell I. (2013). Aquaculture: It's not all about Atlantic Salmon. *Journal of Fisheries and Livestock Production,* 1. doi: https://doi.org/10.4172/2332-2608.1000e103. Accessed 1 June 2023.

Cho, R. (2018). Can soil help combat climate change? Columbia climate school. https://www.clientearth.org/latest/latest-updates/stories/what-is-a-carbon-sink/#:~:text=A%20carbon%20sink%20absorbs%20carbon,oil%2C%20deforestation%20and%20volcanic%20eruptions. Accessed June 18, 2023.

Climate Action Tracker. (2022). State of climate action 2022. https://climateactiontracker.org/documents/1083/2022-10-26_StateOfClimateAction2022_kR0sbBZ.pdf. Accessed June 3, 2023.

Cline, W. (2007). *Global warming and agriculture.* The Center for Global Development and the Peterson Institute for International Economics.

FAO. (2015). El Nino. https://www.fao.org/el-nino/en/. Accessed June 13, 2023.

FAO. (n.d.). Disaster Risk Programme to strengthen resilience in the Dry Corridor in Central America. https://www.fao.org/fileadmin/user_upload/emergencies/docs/Corredor_Seco_Breve_EN.pdf. Accessed June 13, 2023.

Global Forest Watch/World Resources Institute. (2023). Summary. https://www.globalforestwatch.org/dashboards/. Accessed June 27, 2023.

Gulf of Maine Research Institute. (2023a). Gulf of Maine Warming Update: 2022 the Second-Hottest Year on Record. https://www.gmri.org/stories/warming-22/. Accessed 28 June 2023.

Gulf of Maine Research Institute. (2023b). Gulf of Maine Warming Update: Winter 2022–2023. https://gmri.org/stories/gulf-of-maine-warming-update-winter-202223/. Accessed 28 June 2023.

Hodgson, C., Bernard, S., & Harris, B. (2023). Tropical forests shrunk by 10% in 2022. *Financial Times*, June 27. https://www.ft.com/content/1ad02550-f5db-4ba1-a401-23066a1dd0a1. Accessed June 27, 2023.

Intergovernmental Panel on Climate Change (IPCC). (2019). Special Report on the Ocean and Cryosphere in a Changing Climate: Summary for Policymakers. Accessed at www.ipcc.ch/srocc/ on December 16, 2022.

Intergovernmental Panel on Climate Change (IPCC). (2021a). *Climate change 2021: The physical science basis.* Contribution of working group I to the sixth assessment report of the intergovernmental panel on climate change. https://www.ipcc.ch/report/sixth-assessment-report-working-group-i/. Accessed June 15, 2022.

IPCC. (2021b). Technical summary. In *Climate change 2021: The physical science basis.* Contribution of working group I to the sixth assessment report of the intergovernmental panel on climate change. Cambridge University Press, Cambridge.

IPCC. (2022). *Climate change 2022: Impacts, adaptation and vulnerability.* Contribution of working group II to the sixth assessment report of the intergovernmental panel on climate change. https://www.ipcc.ch/report/ar6/wg2/downloads/report/IPCC_AR6_WGII_Chapter05.pdf. Accessed June 15, 2023.

OECD. (2022). Soil carbon sequestration by agriculture. https://www.oecd-ilibrary.org/docserver/63ef3841-en.pdf?expires=1687183802&id=id&accname=guest&checksum=B4878B903C111DF66693E8030FB81D8A. Accessed June 19, 2022.

Rosa, L., & Gabrielli, P. (2023). Achieving net-zero emissions in agriculture: a review. *Environmental Research Letters, 18*(6). https://doi.org/10.1088/1748-9326/acd5e8. Accessed October 13, 2023.

UN Climate Change News. (2023). Healthy forests, healthy planet, healthy humans. https://news.un.org/en/story/2023/03/1134677#:~:text=The%20UN%20established%20the%20Decade,protect%20illegal%20fauna%20trafficking%20in. Accessed October 13, 2023.

United Nations. (2023). Climate change. Land use, land-use change and forestry (LULUCF). https://unfccc.int/topics/land-use/workstreams/land-use--land-use-change-and-forestry-lulucf. Accessed June 15, 2023.

United Nations Environment Programme. (2022). Spreading like wildfire. https://unfccc.int/news/un-scales-up-climate-action-to-protect-forests#:~:text=Twelve%20countries%20announced%20in%20the,and%20land%20degradation%20by%202030. Accessed June 27, 2023.

US NOAA. (2023a). Quantifying the ocean carbon sink. https://www.ncei.noaa.gov/news/quantifying-ocean-carbon-sink. Accessed June 13, 2023.

US NOAA. (2023b). News around NOAA. National program. NOAA declares the arrival of El Nino. https://www.weather.gov/news/230706-ElNino#:~:text=El%20Nino%20is%20a%20natural,far%20beyond%20the%20Pacific%20Ocean. Accessed June 13, 2023.

Voosen, P. (2020) Seas are rising faster than ever, *Science*, November 18. https://www.sciencemag.org/news/2020/11/seas-are-rising-faster-ever. Accessed 26 January 2021.

Waring, B., et al. (2020). Forests and decarbonization—Roles of natural and planted forests. *Frontiers in Forests and Global Change, 3*, 8 May. https://doi.org/10.3389/ffgc.2020.00058. Accessed June 27, 2023.

Woods Hole Oceanographic Institution. (2023). Phytoplankton. https://www.whoi.edu/know-your-ocean/ocean-topics/ocean-life/ocean-plants/phytoplankton/. Accessed June 25, 2023.

WMO. (2023). WMO update: Prepare for El Niño. https://public.wmo.int/en/media/press-release/wmo-update-prepare-el-ni%C3%B1o. Accessed June 13, 2023.

Keeping Up with Future Developments

EU Copernicus. https://www.copernicus.eu/
Food and Agriculture Organization (FAO). https://www.fao.org/
US National Oceanic and Atmospheric Agency (NOAA). https://www.weather.gov/news/
World Metrological Organization (WMO). https://www.wmo.int

Finance 11

11.1 Introduction

Climate change poses issues concerning both financial risks and returns. In the insurance industry, the risks from property damage in hurricane prone areas have been obvious for decades, but there also opportunities for some increased returns. There are investment risks and opportunities in many other sectors. For instance, floods are an obvious threat to agriculture, manufacturing and other industries, but at the same time they increase the need for adaptive construction projects. Because government regulations are widespread for insurance and investment, there are also many public policy issues.

The insurance and investment industries pose a unique set of core issues as their responses to climate change issues in the early 2020s have been evolving. In the insurance industry, many firms have reduced their exposure to increasing climate change risks by withdrawing from regional markets where damages from hurricanes, floods and wildfires have increased. These trends have been particularly pronounced in the US in Florida and California, which had been large and growing markets for property damage insurance. While the insurance firms are mitigating their own exposure to climate change risks, they are also reducing individuals' opportunities to mitigate their exposure to climate change risks. An important consequence of this private sector reduction of coverage is a government policy issue: Will governments increase their subsidized emergency insurance programs?

Among private sector banking and investment industry firms, there have been legal and political complications—again especially in the US—that have led many firms to withdraw from agreements to reduce their investments in firms and industries that have not been responsive to calls for them to reduce their emissions; the fossil fuel industry has been a conspicuous target in this regard. Thus, there has been a reduction in mitigation pressures. These issues are in addition to the "greenwashing" issues in the industry and the prospect of the "stranded assets" of their investments, for instance in the fossil fuel industry.

T. Brewer, *Climate Change*, https://doi.org/10.1007/978-3-031-42906-4_11

Section 11.2 focuses on the insurance industry, and Sect. 11.3 addresses banking and investment issues. Both Sects. 11.2 and 11.3 discuss government policy issues. Section 11.4 summarizes the chapter.

11.2 Insurance

Two re-insurance firms—Swiss Re and Munich Re—have been actively alerting clients and publics to the risks posed by climate change for decades. The former "has been analyzing the effects [of] global warming" for "five decades" (Munich Re, 2023). The latter "identified the threat of climate change" in 1979 (Swiss Re, 2023).

11.2.1 Insurance Exposures to Climate Related Risks

Three findings from a 2019 global survey of insurance companies reported by Cho (2022) highlight issues confronting the industry: 72% believed that "climate change will affect their business, but 80% of them have not taken significant steps to lessen climate risks….[As a group] they have $582 billion invested in fossil fuel investments that could be devalued as climate risks increase."

However, the types of climate-related exposure faced by the insurance industry vary among countries and among regions within countries, and the companies' responses to the risks also vary. Furthermore, the policy responses by government agencies at all levels have also varied. It is therefore important to focus on experiences to date to understand what the implications of climate change so far have been for insurers, their customers, areas of especially high risk, and government responses.

There are readily available factual details for the United States because of the increasingly frequent and intensive large-scale climate-related disasters. There is concern about the wider implications for the future of the insurance industry, as well as the social and economic conditions of large regions of the country. A few facts from the recent years of extreme weather events in the US are suggestive of the magnitudes of the consequences (excerpted by the author from Cho, 2022):

- The economic costs of natural disaster in the US during five years to 2022 was $788.4 billion.
- By November 2022, there were 15 events during the year with more than $1 billion each in loses, including droughts and wildfires, severe storms, floods, and hurricanes in different states and regions. The average number of such events in the previous four decades was only between 7 and 8.
- Insured costs from Hurricane Ian were in the range of $53 to $74 billion, and another $10 billion from flooding.

Such events and the long-term trends they reflect have increased interest in insurance industry responses and the implications for individuals and communities in many regions of the country, as illustrated by the items in Box 11.1.

Box 11.1: Synopsis of Insurance Industry Issues posed by Increased Natural Disasters in the US

Private insurance policies for homeowners typically do not cover flood insurance. In Florida, as many as 80% of the flood policies were with the National Flood Insurance Program of the National Emergency Management Agency (FEMA). Claims from hurricanes left FEMA more than $20 billion in debt, based on its borrowing from other parts of the government.

Many large private property insurance companies have left Florida, California and other states. Because of the wildfires in California during 2017–2018, more than 200,000 property insurance policies were not renewed in 2019.

Insurance rates have increased significantly—63% in the Fair Access to Insurance Requirements (FAIR) plan in Louisiana.

Some insurance companies and some state governments—which are important government regulatory agencies in the US federal political system—have developed incentive programs for property owners who add protective measures to their properties—for instance in Florida for making roofs, doors, windows more storm-resistant, and in California for fire-proofing measures.

Source: Excerpted by the author from Cho (2022).

It should also be noted that the social and economic consequences extend beyond the insurance industry and their clients. There are declining prices in some real estate markets, and the broader economies of local communities, regions and even states can be affected.

Yet, it is also true that socio-economic growth has exacerbated the problem because there are more properties of more value in coastal areas, for instance, as a part of a long-term trend of relatively high population growth rates in those areas. The same is true of the population and economic growth in regions at the intersection of large urban areas and forested areas, where wildfires are a danger.

What are the solutions? One possibility is that internal migration patterns will be reversed by people deciding to leave from the risky areas because of a combination of economic and other personal factors. Another is that the marginal financial incentives offered by insurance companies and local, state and national government programs will induce people to upgrade the properties' protections against extreme weather events. Finally, of course, there are adaptation infrastructure measures that governments at all levels can undertake in order to reduce the damages when extreme weather events do occur. The increasing frequency and severity of such events in their many different manifestations, however, will make such measures increasingly expensive and/or less effective.

11.2.2 Environmental, Social, and Governance Issues (ESG)

The insurance industry has become embroiled in ESG issues—i.e. Environmental, Social, and Governance Issues—with pressure from some groups to do more and others to do less. A particularly important one of the former group is the Net Zero Insurance Alliance (NZIA), which was convened by the UN with the objective of encouraging insurance and reinsurance firms to reduce their portfolios to net-zero greenhouse gas emissions. It was created in 2021 and is part of a broader Glasgow Financial Alliance for Net Zero (GFANZ) established after the Glasgow COP 26 meetings. In January 2023, the NZIA announced a protocol with targets for different types of emissions and a date-specific process for implementation of the provisions of the protocol. By March the NZIA had about 30 members, including many of the world's largest insurance and reinsurance companies (UN Environment Programme, 2023a, 2023b).

Then, 23 US state Attorneys General sent a letter to NZIA members, warning them that they may be violating US anti-trust laws. Although there had been awareness of the issue within NZIA, there were reasons to think there were legal arguments that could successfully counter an anti-trust case. Nevertheless, many major insurance and reinsurance firms from countries in Europe and Asia as well as the United States, quickly withdrew their membership from the NZIA. There were many short-term and long-term implications for decisions by individual companies in many countries, developments in US state and national courts, and other international agreements, as of the end of June 2023.

In Europe, EU competition policy is more stringent than the US anti-trust policies; yet, the EU has explicitly declared that the ESG alliances do not violate EU competition policy. At the same time, the EU has been actively opposing and exposing instances of ESG greenwashing (as discussed below in Sect. 11.3.2).

Insurance companies have also encountered issues about the implications for their future incomes of "stranded assets" and whether their claims about adhering to emissions reduction standards are always accurate. Both of these issues are addressed in the next section in regard to the many sub-sectors of the finance sector.

11.3 Investment

Other participants in the Glasgow Financial Alliance for Net Zero (GFANZ) face some of the same issues, as discussed in the previous section. The GFANZ has more than 550 member companies from seven sub-sectors of the finance sector. In addition to insurance, they are banks, asset owners, asset managers, financial service providers, and investment consultants. Of particular interest to these many sub-sectors are "stranded assets" and "greenwashing."

11.3.1 Stranded Assets

Stranded Assets pose issues of increasing relevance to the many kinds of participants in the financing of business. The issues are particularly salient in *fossil fuels,* including the production and distribution infrastructures, *industries* such as cement, steel, and chemicals that are dependent on fossil fuel inputs in their production processes, and *buildings* because of their dependence on fossil fuels for fossil fuels in their heating systems and air conditioning systems that use electricity from fossil fuel power plants. The financial stakes are enormous for owners, other investors, communities whose economies are dependent on fossil fuels directly or indirectly, as well as local, regional and national governments whose revenues are dependent on these economic sectors.

This section of the chapter thus addresses the following questions about stranded assets: What does the term mean? Why are some economic sectors especially vulnerable? What amounts of money are at stake? Who are the people who are most likely to lose money? What are the technological, economic and political processes that could lead to specific instances of stranded assets? What are individuals, organizations and governments doing to reduce the risks?

The analysis here draws upon several kinds of available studies: introductions to key concepts and questions, e.g. wide-ranging, detailed analysis of many issues, (Caldecott et al. 2016); periodic reviews of the issues and studies; projections based on empirical-mathematical models; and focused analyses on specific kinds of facilities (Energy Monitor, 2021; Fofrich, 2020).

Meaning of the term

The basic concept is that the physical and financial assets of companies—and in some cases governments—will be diminished in the future because of a decline in their expected value in view of climate-related changes in their technological, economic, and/or political situations. A more formalistic financial theory definition that focuses on the process of "Asset Stranding" is: "the collapsing expectations of future profits from invested capital (the asset) as a result of disruptive policy and/or technological change.... Asset stranding becomes a social concern where these effects destabilize markets with negative repercussions in the real economy such as on pensions and government finances" (Semieniuk, 2022a, 2022b). Thus, there are not only direct potential consequences for those with ownership and other stakes in the corporations, there are also potentially larger macro-economic, political, and social consequences.

Vulnerable economic sectors

Some economic sectors are more vulnerable than others. Fossil fuels and the related infrastructure in the coal, oil, and natural gas industries are commonly noted to be especially vulnerable. In addition, as noted above, in the industry sector, the chemical, steel and chemical industries are vulnerable.

Money at stake

It has been estimated that US$1.4 trillion may be at stake in global oil and gas assets (Semieniuk, 2022a, 2022b). However, coal is the most vulnerable in terms of the relative magnitude compared with its present value.

People who are likely to lose money

There is a wide range of potential financial losers in cases of stranded assets—ranging from individual employees of the affected firm to suppliers of the firm to school systems and other public services in surrounding communities to regional and national economic economies. Two questions of special interest in financial analyses are: Who are the investors that would lose money? Could there be a systemic financial crisis?

An extensive study by Semieniuk, (2022a, 2022b) concluded that over half of the losses would be suffered by private persons, especially in the US and Europe. The financial costs of the physical stranding of facilities would be greatest in the US and Russia, followed by China and Canada. There would also be direct losses to government owners of the stranded assets.

As for the possibility of a financial system crisis, financial sector losses are likely to be the largest in the US and UK, compared with other countries. Since both of these countries' financial systems are highly internationalized in many ways, one might imagine that a national systemic crisis in either country could have widespread international consequences—a topic that will surely attract increasing attention among government regulatory agencies.

Technological, economic and political processes that could lead to specific instances of stranded assets

The emergence of solar and wind technologies as economically competitive after large declines in their costs are examples of technological-economic changes that can threaten to make fossil fuel assets less valuable. Alternatively, changes in government policies concerning subsidies for solar and wind can also have a similar effect. In combination, of course, these developments can make coal mines and infrastructure in particular vulnerable to becoming stranded assets.

Reducing the risks

There are many ways that stranded asset losses can be reduced—including actions already in progress. Some fossil fuel companies have begun to diversify their operations, and of course some banks and other financial services firms have reduced their fossil fuel investments as a result of ESG commitments. Yet, ESG agreements and the pledges made by firms have become topics of widespread controversy, as noted above in Sect. 11.2.2 concerning the insurance industry and also below in the following section on "greenwashing."

11.3.2 Greenwashing

The term "greenwashing" refers to misrepresentations by business of their policies and actions in order to appear more climate-friendly than they actually are. It is an issue of increasing salience and significance as extreme weather events and the projections of the future in climate change research reports create an increasing sense of urgency about the need for more action.

Interactions among the many organizations involved in "greenwashing" issues are complex, contentious, and evolving. The organizations include:

- National and international NGOs that have been supporting the establishment of sustainability standards and the monitoring of industry compliance with the standards.
- Governments and international agencies that have been developing codes of conduct—some of which impose legal obligations on firms while others are hortatory and without sanctions for non-compliance.
- Industry associations that have exhibited a variety of responses to the issues—some of them supportive at least in principle, others focused on delaying the development and applications of standards.
- Individual firms in many economic sectors—among which there has been much variation in attitudes and actions.
- Intense interest in the issues among financial services firms, including retirement funds and other investment funds and advisory services.

Much of the attention to the issues focuses on the substance of the many codes and government regulations, transparency in whether firms adhere to them, and what to do in the case of violations.

Some codes are intended to be generally applicable to firms in all sectors; others are industry-specific; and yet others are issue-specific and multi-sectoral. The EU has been especially prominent in its concerns about greenwashing by banks and other financial firms (just as is has been prominent in its *support for ESG* as indicated in Sect. 11.2.2 above in regards to the insurance industry). EU regulation of financial sector firms is institutionalized in three European Supervisory Authorities (ESAs): European Banking Authority (EBA); European Insurance and Occupational Pensions Authority (EIOPA); and European Securities and Markets Authority (ESMA). In June 2023 they coordinated the release of their three individual reports about greenwashing (EBA, 2023a, 2023b; EIOPA, 2023; ESMA, 2023; also see Kenza & Hancock, 2023; and RepRisk, 2023). Box 11.2 highlights the EU authorities' findings.

Box 11.2: Highlights from EU Supervisory Authorities' Reports about Greenwashing

All sectors (EBA)

As a whole (i.e., considering all companies and sectors), the total number of alleged cases of misleading communication on ESG related topics reported by stakeholders has risen significantly in the recent years. It has been multiplied by 4 since 2018 and 6.5 since 2012. This rise has occurred in all regions, but it has been especially high for companies located in North America and in the EU, which accounted for 60% of all alleged cases of greenwashing in 2022.

Alleged cases of greenwashing have also been occurring in all economic sectors. However, it has been mostly concentrated around six activities including oil, gas and utilities, mining, industrial construction, food and beverage, household goods and the financial sector. The latter accounted for ca. 16% of alleged greenwashing cases observed worldwide in 2022, including insurance (1%) banks (4%) and financial services (10%).

Securities and Markets (ESMA)

Misleading claims may relate to all key aspects of the sustainability profile of a product or an entity such as ESG governance and resources; ESG strategy, policies and credentials; ESG performance metrics and targets; and sustainability impact. Cherry-picking, omission, ambiguity, empty claims (including exaggeration), misleading use of ESG terminology such as naming and irrelevance, are seen as most widespread misleading qualities.

A total of 138 respondents participated [in a survey] spanning across 20 EU Member States in addition to non-EU countries (17 respondents). Respondents were asked to identify the risk of each channel serving to communicate misleading sustainability claims made at entity level and/or at product/service level. Marketing materials came out significantly in the lead in this respect (73%). On the other hand, regulatory documents were perceived as the channel carrying the least risk for greenwashing (43%). Respondents were invited to provide what they considered as examples of potential greenwashing practices. A total of more than 80 examples were provided by 50 respondents.

Banking (EBA)

Examples of greenwashing in the EU banking sector indicate that a bank can potentially engage in greenwashing in multiple ways, mostly at entity level….

The past years data related to misleading communication on ESG topics shows a clear increase in the total number of potential cases of greenwashing across all sectors including EU banks.

Insurance and Pensions (EIOPA)

EIOPA finds that greenwashing can manifest—to varying extents—as part of the broader set of conduct risks at all stages of the insurance (e.g., entity level, product manufacturing, delivery and management) and pensions (e.g., scheme design, delivery and management) lifecycles. Throughout the different stages highlighted in the report, potential examples of greenwashing are included to show how greenwashing can occur in practice.

The term "misleading" is understood in this report as an umbrella term that covers selective disclosure, empty claims, omission, lack of disclosure, vagueness, lack of clarity, inconsistency, lack of meaningful comparisons, unsubstantiated underlying assumptions, misleading imagery, irrelevance, outdated information, and falsehoods.

Sources: Excerpted by the author from EBA (2023a; 2023b), EIOPA (2023), ESMA (2023).

11.4 Conclusion

The finance sector and climate change issues interact so that each affects the other in many ways. The chapter begins with the effects on the insurance industry of the damage done to property and lives by hurricanes and other storms, floods, droughts, and fires. The effects are now felt in all regions of the world and not only in coastal areas. The impact of large and increasing losses and thus claims on insurance companies have led them to reduce their coverage in many regional markets. This, however, creates an opposite cause-effect relationship such that decreases in insurance coverage expose populations to greater un-insured losses. In some instances, governments have imposed new regulations limiting insurance companies' market retreats, and government have also developed government-sponsored insurance programs to offset private insurers' reductions of coverage.

Another issue that has affected the insurance industry is the establishment of ESG programs whereby insurance companies—and other kinds of financial services firms—are encouraged to reduce the exposure of their portfolios of policies and investments in firms and sectors that are carbon-intensive emitters. These arrangements have been targeted by some US states and politicians as contrary to US anti-trust laws. How this controversy will evolve in legal and political institutions in the US is not yet clear. At the same time, in Europe the EU has taken the opposite approach by preemptive declarations that ESG programs do not violate European competition policies. These ESG issues are also evident in other financial services industries.

Two other issues that are evident in investment services are "Stranded Assets" and "Greenwashing." The former pertain especially, but not only, to fossil fuel assets. The basic concept is that many kinds of physical and financial assets can become less valuable because of technological, economic or policy changes that occur as part of processes that mitigate carbon emissions. Estimates of the magnitude of potential losses are on the order of US$ trillions.

Greenwashing is yet another issue that has embroiled the finance sector. Greenwashing occurs when a firm misrepresents its intentions, policies, plans or actions in regard to reducing its carbon footprint directly or indirectly. The issue is important in the finance sector both because firms in the sector may engage in greenwashing and because financial firms that invest in other firms in other sectors as well as the finance sector may find their investments are sometimes in firms that make deceptive or outright false claims about their climate related behavior.

In short, climate change poses centrally important issues to private and public participants in finance sector business, and such participants' behavior poses important issues for the future of climate change.

Questions to Ponder

1. What "greenwashing" standards, monitoring and reporting do you think should be available? Who should develop and enforce them?
2. Do you think governments should insure coastal property against storms and floods where private insurance companies will not? If yes, what kinds of financial risk and return policies should the government insurance agencies adopt?
3. Which industries face what kind of "stranded asset" issues?
4. What legal issues are involved in agreements among financial services firms to screen their investments on the basis of sustainability criteria of potential investments or actual investments? What are the international variations among jurisdictions in the issues?

References

Bryan, K., & Hancock, A. (2023). EU regulators flag rising greenwashing practices by banks. https://www.ft.com/content/5d236244-e073-412d-b981-0d2757f60b4b. Accessed June 7, 2023.

Caldecott, B., et al. (2016). *Stranded assets: A climate risk challenge*. Inter-American Development Bank. Accessed June 7, 2023.

Cho, R. (2022). State of the Planet. With Climate Impacts Growing, Insurance Companies Face Big Challenges. https://news.climate.columbia.edu/2022/11/03/with-climate-impacts-growing-insurance-companies-face-big-challenges/. Accessed June 6, 2023.

Climate Champions. (2023). Mobilizing finance to accelerate climate action and advance the SDGs. https://climatechampions.unfccc.int/mobilizing-finance-to-accelerate-climate-action-and-advance-the-sdgs/. Accessed June 7, 2023.

Economist. (2023a). Who is keeping coal alive? https://www.economist.com/search?q=who+is+keeping+coal+alive. Accessed June 7, 2023.

Economist. (2023b). What would the perfect climate-change lender look like? https://www.economist.com/search?q=What+would+the+perfect+climate-change+lender+look+like. Accessed June 7, 2023.

Energy Monitor. (2021). Why airports could become stranded assets. https://www.energymonitor.ai/sectors/transport/why-airports-could-become-stranded-assets/. Accessed June 7, 2023.

European Banking Authority (EBA). (2023a). ESAs present common understanding of greenwashing and warn on related risks. https://www.eba.europa.eu/esas-present-common-understanding-greenwashing-and-warn-related-risks. Accessed June 10, 2023.

European Banking Authority (EBA). (2023b). EBA Progress Report on Greenwashing Monitoring and Supervision. https://www.eba.europa.eu/sites/default/documents/files/document_library/Publications/Reports/2023/1055934/EBA%20progress%20report%20on%20greewnwashing.pdf. Accessed June 10, 2023.

European Securities and Markets Authority (ESMA). (2023). Progress Report on Greenwashing. https://www.esma.europa.eu/sites/default/files/2023-06/ESMA30-1668416927-2498_Progress_Report_ESMA_response_to_COM_RfI_on_greenwashing_risks.pdf. Accessed June 10, 2023.

European Insurance and Occupational Pensions Authority (EIOPA). (2023). Advice to the European Commission on Greenwashing. https://www.eiopa.europa.eu/system/files/2023-06/EIOPA%20Progress%20Report%20on%20Greenwashing.pdf. Accessed June 10, 2023.

Feingold, S. (2022). Norway's massive sovereign-wealth fund sets net-zero goals. https://www.weforum.org/agenda/2022/09/norways-massive-sovereign-wealth-fund-sets-net-zero-goal/. Accessed May 26, 2023.

Robert Fofrich, R. (2020). Early retirement of power plants in climate mitigation scenarios. *Environmental Research Letters, 15*, 9. https://iopscience.iop.org/article/10.1088/1748-9326/ab96d3/meta. Accessed June 7, 2023.

Inman, P. (2023). Green investment funds pushing money into fossil fuel firms, research finds. https://www.theguardian.com/business/2023/may/02/green-investment-funds-pushing-money-into-fossil-fuel-firms-research-finds. Accessed June 7, 2023.

Kenza, B., & Hancock, A. (2023). EU regulators flag rising greenwashing practices by banks. *Financial Times,* 1 June 2023. https://www.ft.com/content/5d236244-e073-412d-b981-0d2757f60b4b. Accessed June 7, 2023.

Matikainen, S., & Soubeyran, E. (2022). What are stranded assets? London School of Economics and Political Science. Grantham Research Institute on Climate Change and the Environment. https://www.lse.ac.uk/granthaminstitute/explainers/what-are-stranded-assets/. Accessed June 7, 2023.

Milne, R. (2023). Norway's oil fund sides with climate activists against ExxonMobil and Chevron. *Financial Times*, 26 May. https://www.ft.com/content/ad012975-0c77-4aba-b0b5-4174290dd2a7. Accessed May 26, 2023.

Mooney, A., & Williams, A. (2023). Nuns urge Citigroup to rethink financing of fossil fuel projects. https://www.ft.com/content/ed0ce147-6308-4141-85c4-ab2fde65c428. Accessed June 7, 2023.

Mundy, S., & Kenza, B. (2023). Inside the fossil fuel stand-off that's ensnared global banks. https://www.ft.com/search?sort=relevance&q=simon+mundy+and+kenza+bryan+inside+the+fossil+fuel+stand-off. Accessed June 7, 2023.

Munich Re. (2023a). Climate change and its consequences. https://www.munichre.com/en/risks/climate-change.html. Accessed May 21, 2023.

Norges Bank Investment Management. (2023). The Government Pension Fund Global. About the Fund. https://www.nbim.no/en/the-fund/about-the-fund/. Accessed May 26, 2023.

Quinson, T. (2023a). What's the Legal Definition of Greenwashing? https://www.bloomberg.com/news/articles/2023-01-11/what-s-the-legal-definition-of-greenwashing-green-insight?leadSource=uverify%20wall. Accessed June 7, 2023.

Quinson, T. (2023b). Soros fund presses ahead despite hurdles. Bloomberg Green. ESG Investing.

RepRisk. (2023). RepRisk in a Nutshell. https://www.reprisk.com/. Accessed June 10, 2023.

Smith, I., & Kenza, B. (2023). Insurance industry turmoil over climate alliance exodus. https://www.ft.com/content/1dd66ce1-a720-4c56-96d9-8d47f07f376f. Accessed June 7, 2023.

Semieniuk, G. (2022a). Stranded fossil-fuel assets translate to major losses for investors in advanced economies. *Nature Climate Change*, 12, 532–538. https://www.nature.com/articles/s41558-022-01356-y#citeas. Accessed June 7, 2023.

Semieniuk, G. (2022b). Stranded fossil-fuel assets translate to major losses for investors in advanced economies. *Nature Climate Change*, *12*, 532–538. https://www.nature.com/articles/s41558-022-01356-y#citeas. Accessed June 7, 2023.

Surminski, S. (2020). Climate change and the insurance industry: Managing risk in a risky time. *Georgetown Journal of International Affairs*. June 9. https://gjia.georgetown.edu/2020/06/09/climate-change-and-the-insurance-industry-managing-risk-in-a-risky-time/. Accessed June 7, 2023.

Swiss Re. (2023b). Climate Change Risk. https://www.swissre.com/risk-knowledge/mitigating-climate-risk.html. Accessed May 21, 2023.

UN Environment Programme. (2023a). Net Zero Insurance Alliance. https://www.unepfi.org/net-zero-insurance/. Accessed June 7, 2023.

UN Environment Programme. (2023b). World-Leading Insurers and United Nations Launch Pioneering Target-Setting Protocol to Accelerate Transition to New-Zero Economy. https://www.unepfi.org/industries/insurance/launch-of-nzia-target-setting-protocol-version-1-0/. Accessed June 7, 2023.

World Economic Forum. (2023). Norway's massive sovereign-wealth fund sets net-zero goals. https://www.weforum.org/agenda/2022/09/norways-massive-sovereign-wealth-fund-sets-net-zero-goal/. Accessed June 7, 2023.

Resources for Tracking Future Developments

EDHEC Risk Climate Impact Institute. https://climateimpact.edhec.edu/edhec-risk-climate-impact-institute-publications

Munich Re. Climate change and its consequences. https://www.munichre.com/en/risks/climate-change.html

OECD. https://www.oecd.org/cgfi/forum/

Swiss Re. Climate Change Risk. https://www.swissre.com/

UN Environment Programme. Net Zero Insurance Alliance. https://www.unepfi.org/net-zero-insurance/

Part IV
The Future

12 Climate Model Projections and Potential Action Paths

If we continue on the business as usual trajectory,
there will be a tipping point that we cannot avert....
We will indeed drive the car over the cliff.
Holdren (2009)

The climate time-bomb is ticking.... The 1.5C limit is achievable.
But it will take a quantum leap in climate action.
Guterres (2022)

12.1 Introduction

The header quotes for the chapter are apt summaries of the themes of this book, as well as the state of the world. As the first quote states, the reality is that the *problems* of climate change facing the world are existential. The second quote emphasizes the urgent need for *action*. Indeed, an IPCC Summary for Policy-makers based on hundreds of studies using scientific evidence and mathematical models to make projections of the future concluded that:

> All global modelled pathways that limit warming to 1.5 °C ... with no or limited overshoot, and those that limit warming to 2 °C ..., involve *rapid* and *deep* and in most cases *immediate* GHG emission reductions *in all sectors* (IPCC, 2022: 24; italics added).

T. Brewer, *Climate Change*, https://doi.org/10.1007/978-3-031-42906-4_12

This chapter thus focuses on the problems, as revealed by climate science, and the choices of paths of action into the future that are available to solve the problems. The themes are that:

The core results of climate science research are clear and conclusive. There are many types of evidence from numerous sources that the problems are increasing in frequency and intensity. Some of the problems are already occurring, and projections of the future are that they will become worse. It is thus common for climate scientists and other experts to refer to the present situation as a "climate emergency" or "existential crisis."

Ethical issues about climate change are also increasingly salient. They include a mix of value-based, fact-based, and logical analyses that draw upon established concepts and principles from ethical traditions. There is a consensus that producers and consumers of the sources of the gasses and particulate matters that cause climate change are obligated on ethical grounds to adopt more stringent mitigation measures and provide financial assistance for adaption measures. These obligations exist for societies' collective actions at all levels as well as for all individuals' actions.

Technological, economic and political paths that include diverse mitigation and adaptation actions are in progress. Although there is variety in their stages of technological development, economic cost-effectiveness, economic scale, and political feasibility, they offer hope that more and more of the necessary actions will be undertaken.

The sections of the chapter discuss: projections of temperatures; melting ice in the Arctic and Antarctic regions and their global effects; climate tipping points; investment-and-economic growth tipping points; the *tragedy of the commons* and *market failures*; and the conclusion. Four annexes provide details about: climate change models; sector-specific policies and priorities for action; systemic shocks; and new data about the earth's carbon budget.

12.2 Projections of Temperatures

Studies of climate futures are developed to a substantial extent on the basis of empirically based studies of the past, as presented in the preceding chapters of this book. The studies are historically grounded in facts about decades, centuries and even millennia of changes in temperatures, and their causes and consequences. The studies also include mathematical relationships among variable features of the climate that can be projected in the context of multiple assumptions about the future. The studies thus reflect a wide range of imaginable technological, economic, political and other kinds of developments that can alter projections of temperatures.

Periodic IPCC reviews of projections of the future—together with many other reports—are essential resources for any serious consideration of what climate conditions and their impacts will be. In the Summary for Policymakers in IPCC

Assessment Report 6 (2021a: 14, B.1), the following succinct statement is an appropriate starting point for considering the projections:

> Global surface temperature will continue to increase until at least mid-century under all emissions scenarios considered. Global warming of 1.5 °C and 2 °C will be exceeded during the 21st century unless deep reductions in CO_2 and other greenhouse gas emissions occur in the coming decades.

A subsequent report by the IPCC (2023: B.1; B.1.2) concluded specifically about the near-period:

> Global warming will continue to increase in the near term (2021–2040) mainly due to increased cumulative CO_2 emissions in nearly all considered scenarios and modelled pathways. In the near term, global warming is more likely than not to reach 1.5 °C even under the very low GHG emission scenario (SSP1-1.9) and likely or very likely to exceed 1.5 °C under higher emissions scenarios.
>
> Continued emissions will further affect all major climate system components. With every additional increment of global warming, changes in extremes continue to become larger.

More detail about the temperature projections of five models for short-term, medium-term and long-term time periods are presented in Table 12.1. The projections are comparisons with the global mean surface temperature during 1850–1900.

These projections can also be depicted graphically, as in Fig. 12.1. Note that the average for the 1850–1900 period is the "base line" from which the changes

Table 12.1 Model projections of future changes in mean global surface temperatures

Models[a]	Scenarios of Emission Levels[b]	Best estimate of increase in mean global surface temperature compared with 1850–1900 mean (°C)		
		Near-term (2021–2040)	Mid-term (2041–2060)	Long-term (2061–2100)
SSP1-8.5	Very high	1.6	2.4	4.4
SSP1-7.0	High	1.5	2.1	3.6
SSP1-4.5	Intermediate	1.5	2.0	2.7
SSP1-2.6	Low	1.5	1.7	1.8
SSP1-1.9	Very low	1.5	1.6	1.4

Source Excerpted and compiled by author from IPCC (2021a: 14; also see IPCC 2023: Box SPM.1)
[a]The five models in the table are "illustrative emission scenarios" included in IPCC (2021a). See Annex 12.1 of this chapter for more details about the content of the models. Also see Annex 2.2 in Chap. 2, for more information about the methods of measurement
[b]Emission scenarios: very high = emissions double by 2050; high = emissions double by 2100; intermediate = emissions remain at current levels until 2050; low = emissions are net zero after 2050; very low = emissions are net zero in 2050

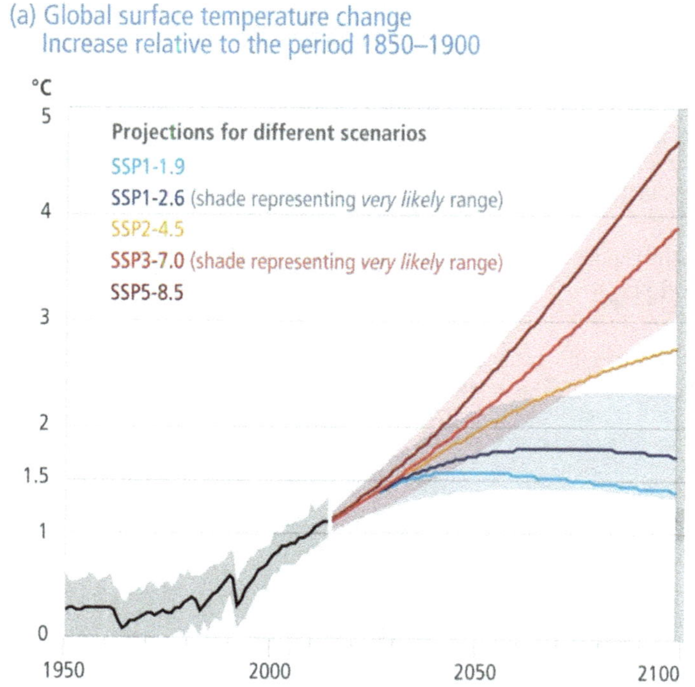

Fig. 12.1 Graphic display of model projections of future changes in mean global surface temperatures. *Source* IPCC (2021b); used with permission; also see IPCC (2023: Fig. SPM.4(a))

have been computed for the 1950–2100 period in the chart. The 1850–1900 period base is 0 degrees C on the chart.

The data in the *middle* projection indicate continuing global temperature *increases* for the rest of the century. In addition, the cumulative results of a large number of studies based on different assumptions, variables and data sets clearly indicate that for decades into the future climate change will become much more problematic—unless there are major changes in, business practices, government policies, and consumer behavior, as well as technologies.

It should also be noted that these are all global averages and that there have been and will be substantial and consequential variations among regions. Regional variations are especially noteworthy for the Arctic and Antarctic, for which expanding data bases and modelling exercises have led to increased concern, as reflected in the following section.

12.3 Melting Ice in the Arctic and Antarctic Regions with Global Consequences

The IPCC has projected that "It is virtually certain that the Arctic will continue to warm more than the global surface temperature, with *high confidence above two times* the rate of global warming" (IPCC, 2021a: 15; italics added). The focus in this chapter is on the *future* and the consequences of the ice melt for sea level rises and ocean currents far beyond the two polar regions. Both regions have been experiencing record-shattering ice melts, with global implications.

In the Arctic region, the Greenland glacial melt poses different issues from the melting ice floating in the Arctic Ocean (Brewer, 2019). A study of the melting glacier on Greenland noted that "future mass loss of the [Greenland Ice Sheet] is challenging to predict because it is a non-linear function of temperature and occurs over long timescales" (Honing et al. 2023; also see American Geophysical Union, 2023). The authors consequently used an Earth System model that incorporates more contextual data than widely-used climate change models (see Annex 12.1). They projected "two critical volume thresholds" for carbon dioxide emissions—one threshold at 1000 gigatonnes (Gts) of carbon dioxide that could lead to the southern portion to melt and the other threshold at 2500 Gts that would lead to a melt of almost all of the ice. As of their research in the early 2020s, the cumulative total was already 500 Gts and thus halfway to the 1000 Gt critical level.

A key question, therefore, concerns the temperature increase above the pre-industrial average that would be a *tipping point*—i.e. the point at which a "mass [ice] loss will inevitably continue until a substantial part of the ice sheet has melted." The next key question concerns the temperature that would cause the mass loss. The answer is "within the range of the temperature limits of the Paris agreement"—i.e. 1.5 and 2.0 °C. As the lead author Honing remarked (American Geophysical Union, 2023): "The first tipping point is not far from today's climate conditions, so we're in danger of crossing it. ... Once we start sliding, we will fall off this cliff and cannot climb back up." (You might want to ponder the first header quote at the beginning of this chapter.)

In the Antarctic, evidence of continuing record-setting annual decline rates of the extent of the sea ice has prompted more worrisome projections of its future contributions to sea level rise (National Snow and Ice Data Center (NSIDC), 2023a). In February 2023, the extent of the Antarctic sea ice reached a record low (NSIDC, 2023b). There is increasing evidence that the meltwater from the Antarctic melting ice is causing a slowdown in the Atlantic Meridional Overturning Circulation (AMOC) (Li et al. 2023; Sinet et al. 2023).

12.3.1 Effects on Sea Level Rise

A satellite-based US NASA system that is coordinated with the French Centre National d'Etudes Spatiales (CNES) collected 30 years of evidence of sea level rise from 1993 to 2022 (NASA & CNES, 2023). That data is based on microwave signals that "bounce" back to the satellites; it is combined with surface-based data from coastal monitoring systems. The data reveal that the sea level rise in 1993 was 0.20 cm (0.08 in) and 0.44 cm (0.17 in) in 2022, thus slightly more than a doubling over the 30 years. An extrapolation to 2050 is 0.66 cm (0.26 in) per year.

Although these may seem like small, inconsequential rises, they are enough to threaten millions of people in highly populated urban areas around the world. Some large coastal urban areas are already experiencing much greater sea level rise than the global averages—for instance, Houston, Miami and New Orleans because of extraordinary rises in the Atlantic Ocean and Gulf of Mexico coasts in the southeast of the US (Dangendorf et al. 2023; Yin, 2023). In addition, there are coastal areas where land-levels are subsiding because of extractions of water from underground aquifers.

The IPCC (2023: B.3.2) has summarized the effects of Arctic and Antarctic ice melts on sea level rise as follows: "At sustained warming levels between 2 and 3 °C, the Greenland and West Antarctic ice sheets will be lost almost completely and irreversibly over multiple millennia, causing several meters of sea level rise."

12.3.2 Effects on Ocean Currents

A more complicated and less visible but nevertheless also consequential effect of the Arctic and Antarctic ice melts is the effect on ocean currents. Greenland's ice melts will have significant long-term effects on Atlantic Ocean currents; this ocean current is known as the Atlantic Meridional Overturning Circulation (AMOC). The AMOC is especially notable because it is marked by its effects on countries on both side of the North Atlantic. Thus, while the emission mitigation issues focus on the sources of the gases and particulate matter from as far as Asia that are causing the Greenland ice sheet to melt, the adaptation issues focus on the coastal regions of North American and Western European countries in addition to people and ecosystems within the Arctic region.

Research on the effects of Antarctic ice melt on the overturning circulation of the oceans has projected a rate as high as twice the rate of the Arctic melt (Li et al. 2023; Sinet et al. 2023). Visual representations of the global ocean conveyor belt, including the AMOC, are available at NSIDC (2023b).

The expanding evidence and concern from projections of future developments—including tipping points in both the Arctic and Antarctic regions—have occurred in the context of additional research on many other types of tipping points. Indeed, both cases of polar ice melt and their effects on sea levels and ocean currents are among a large number of Climate Tipping Points that are topics of intensive research.

12.4 Climate Tipping Points (CTPs)

A Climate Tipping Point (CTP) has been defined as "a critical threshold beyond which a system reorganizes, often abruptly and/or irreversibly" (OECD, 2022: 8; also see IPCC, 2021a). *Recent climate studies have led to a lowering of the projected temperature thresholds and shortening of the time of occurrence of CTPs* (Armstrong McKay et al. 2022; Lenton, 2019; IPCC, 2018, 2021a, 2021b; Klower et al. 2021; Lindwall, 2022; OECD, 2022). The increased understanding has drawn upon information from a combination of paleolithic studies, observational studies, and model-based studies (Armstrong McKay et al. 2022). It is now understood, for instance, that CTPs can occur as a result of identifiable *non-linear* relationships (Alley et al. 2003; Klower et al. 2021).

CTPs are thus now regarded as needing more urgent action than previously recognized. A review of CTP research concludes that:

> While understanding of the risk associated with the crossing of climate system tipping points and the potential for cascading effects has considerably increased over the past decade … mitigation efforts still fall far short of what is needed to avoid the crossing of these thresholds, nor are current adaptation efforts sufficient to manage them (OECD, 2022: 54).

The examples in Box 12.1 reveal the variety of potential CTPs that are under intensive analysis.

Box 12.1: Examples of Projected CTP Thresholds

Permafrost thawing and collapse

There are permafrost CTPs located in or near the Arctic circle, and they pose potentially catastrophic global impacts because of the potential quantities of methane releases—over the short-term and the long-term. Permafrost thawing has already been occurring within the Arctic region for many years. Estimates of CTPs for permafrost *thawing* temperatures are as low as 1.5 °C, but separately much higher for permafrost *collapse* at 4 °C.

Coral reef die-offs

Coral reef dieoff poses issues in the South Pacific Ocean—and different mitigation and adaptation issues. Coral reef dieoff north of Australia is projected to occur at 1.5 °C. However, the lower "minimum estimate" is as low as 0.8 °C, which is a temperature that has already been surpassed.

Forest dieback

The Amazon rainforest that straddles the equator in Brazil and eight other countries and Boreal forests that are in the far northern hemisphere are both vulnerable to massive diebacks. However, the Amazon area tipping point is at relatively higher temperature. Amazon rainforest dieback in Brazil is projected to occur at 3.5 °C, or within a minimum-to-maximum range of 2.0–6.0 °C.

Sources: Armstrong McKay et al. (2022), IPCC (2021a, 2023), OECD (2022).

In terms of the geographic extent of their impacts, CTPs are often classified as being either "global" or "regional":

> At the *regional* level, individual tipping points are associated with different types of potentially severe regional or local impacts, such as extreme temperatures, higher frequency of droughts, forest fires and unprecedented weather. At the *global* scale, tipping points would lead to world-wide impacts through e.g. contributing to additional greenhouse gas emissions into the atmosphere and temperature feedback loops or to faster sea-level rise (OECD, 2022: 8; italics added).

The list of sixteen CTPs in Table 12.2 contains a list of nine "global" and seven "regional" CTPs.

The projected CTPs, especially those with threshold temperatures less than 2 °C, are arguably overdue for more mitigation and adaptation actions. A succinct phrasing of the CTP action issues is: Who could/should do what and when about mitigation and adaptation? The answer to the *when* question is: "immediately or otherwise as soon as possible, but in any case by 2029." Such a message results from studies about potential impacts of CTPs and about *actions* to mitigate emissions and actions to adapt to their potential impacts. An IPCC (2023: C.2.1; italics added) synthesis of studies concluded that:

> *Deep, rapid, and sustained mitigation and accelerated implementation of adaptation actions in this decade* would reduce future losses and damages related to climate change for humans and ecosystems (very high confidence). As adaptation options often have long implementation times, accelerated implementation of adaptation in this decade is important to close adaptation gaps (high confidence).

Answers to the *who* and *what* questions vary among the types of CTPs and their locations, origins and impacts. The following are only illustrative of possible action paths; the list is not intended to be comprehensive. *The topic is one of the most important on the climate change action agendas for governments, businesses, and publics, as well as climate scientists.*

Table 12.3 contains excerpts of mitigation options that have been compiled from an extensive list organized in terms of *economic sectors* in IPCC (2023: Fig. SPM.7); in any case, it is an appropriate agenda for mitigation and adaptation issues generally.

Additional details of sectoral policy options and priorities are available in Annex 12.2.

Table 12.2 Projected climate tipping points (CTPs)

Tipping point
Global
Amazon rain forest
Atlantic Meridional Overturning Circulation (AMOC)
Boreal permafrost (collapse)
East Antarctic ice sheet
East Antarctic sub glacial sea basins
Greenland ice sheet
Labrador Sea convection
West Antarctic ice sheet
Arctic winter sea ice (collapse)
Regional
Barents Sea ice
Boreal forest, northern expansion
Boreal forest, southern expansion
Boreal permafrost (abrupt thaw)
Coral reefs
Mountain glaciers[a]
Sahel

Sources Excerpted and compiled by the author from Armstrong McKay et al. (2022) and OECD (2022)
[a]The "Mountain glaciers" tipping point is in Armstrong McKay et al. (2022); the "Tibetan plateau" tipping point is in OECD (2022)

Table 12.3 Potential mitigation measures

Sectors[a]	Mitigation options[b]
Energy supply	Solar
	Wind
	Reduce methane from coal, oil and gas
	Bioelectricity, including BECCS (bioenergy with carbon capture and storage)
	Geothermal and hydropower
	Nuclear
	Fossil carbon capture and storage (CCS)
Land, water, food	Reduce conversion of natural ecosystems (e.g. improved cropland management)

(continued)

Table 12.3 (continued)

Sectors[a]	Mitigation options[b]
	Carbon sequestration in agriculture
	Ecosystem restoration, afforestation, reforestation
	Shift to sustainable, healthy diets[c]
	Improved sustainable forest management
	Reduce methane and N_2O in agriculture
	Reduce food loss and food waste
Settlements and infrastructure	Efficient buildings
	Fuel efficient vehicles
	Electric vehicles
	Efficient lighting, appliances and equipment
	Public transport and bicycling
	Biofuels for transport
	Efficient shipping and aviation
	Avoid demand for energy services
	On site renewables
Industry and waste	Fuel switching
	Reduce emissions of fluorinated gas
	Energy efficiency
	Material efficiency
	Reduce methane from waste/wastewater
	Construction materials substitution
	Enhanced recycling
	Carbon capture with utilization (CCU) and CCS

Source Excerpted and compiled by the author from IPCC (2023: Fig. SPM.7)

[a]These "sectors" do not correspond precisely to the "sectors" in Chaps. 5–11 of this book. In addition, "enhanced health services" were included in the original source, with examples for "nutrition and diets"

[b]Options are listed as in original source, with largest "potential contribution to net emission reductions, 2030"—*within each sector*—at the top of the sector list and those with the least potential at the bottom of the list

12.5 Another Type of Tipping Point: Investment and Economic Growth Opportunities

Recent research has focused attention on a different kind of Tipping Point, which is defined as a time "when a set of conditions are reached that allow new technologies or practices to out-compete incumbents. After a tipping point is crossed, reinforcing feedback loops take hold that drive self-reinforcing progress, so that greater deployment of the solution encourages even faster deployment" (Stern &

Romani, 2023: Fig. 1; also see Songwe et al. 2022; Systemiq, 2021). This notion is thus about how technological and economic changes enhance the competitive position of a relatively climate-friendly good or service so that it can be more widely deployed.

Therefore, compared with the CTPs in the previous section, investment-and-economic-growth tipping points are fundamentally different in important ways. As Stern and Romani (2023: 10) have emphasized:

> Investment and innovation require optimism about and confidence in future possibilities and a recognition of the necessity of change. These are there and growing. A shared vision is crucial to crafting reality. The coming decades will be decisive for the planet, and investment and innovation embodying new technologies, together with the drive to net-zero will yield *the growth story of the 21st century*.

The two kinds of tipping points are complementary in both conceptual and practical terms. Whereas the tipping points in the previous section are about *problems* becoming more dangerous because of thresholds in climate change processes, the tipping points in this section are about *solutions* becoming more attractive because of thresholds in technological and economic processes. One could refer to the former as "negative" TPs and the latter as "positive" TPs. *The combination of the scientific focus of one and the economic focus of the other is a compelling example of the potential for inter-disciplinary perspectives on climate change issues.*

There are now estimates of the timing of the investment-and-economic-growth Tipping Points in a wide range of sectors, as summarized in Table 12.4. It is notable that a tipping point has already occurred for electricity production and that all of the others are projected to occur by 2030.

Table 12.4 Estimated investment and economic growth tipping points

Sector[a]	Year of estimated investment and economic growth tipping point
Electricity	2018
Light road transport	2024
Trucking	2027
Aviation	2028
Shipping	2030
Building heating	2025
Food and agriculture	2026
Land use change	2030
Steel	2030
Cement	2030
Chemicals	2030

Source Excerpted and compiled by the author from Stern and Romani (2023: Fig. 1), as adopted from Systemiq (2021)

[a]The "sectors" are not precisely the same as the "sectors" in Chaps. 5–11

12.6 Political-Economy: "The Tragedy of the Commons" and "Market Failures"

A political-economy approach that focuses on the benefits and costs of climate action (or inaction) offers another complementary inter-disciplinary perspective, which is essential to understanding core issues of politics and economics that will be central to the future course of climate change problem-solving. The political-economy approach has produced many studies that provide key insights into both problems and solutions. In particular, two closely related concepts—"market failures" and "the tragedy of the commons"—have received much attention from economists and political scientists.

Market failures occur when prices in market-based transactions do not include the costs borne by society that are external to the transactions and not paid for by the participants in the transaction (Keohane & Olmsted, 2016; Salanié, 2000; Stiglitz, 1989; Stern, 2006). When an activity such as producing electricity or driving an automobile creates emissions but the emitter does not pay for the cost to society of the emissions, the external costs represent market failures. It has been said that climate change is "the greatest example of market failure we have ever seen" (Stern 2006: 1).

The concept of the *tragedy of the commons* (Hardin, 1968; Ostrom, 1990) focuses on the vulnerability of the commons—i.e. a public good—to the tragic consequences of the incentives for individuals to use the commons for private gain and thus reduce its availability to others. Individually rational decisions thus lead to collectively *ir*rational outcomes; many individuals rationally pursuing their private interests destroy the common good so that it is no longer available to anybody. The basic phenomenon is well known among environmental economists, and it is widely noted in climate change studies (Stavins, 2013; also 2001, 2019).

Government regulatory policies can reduce the incidence of market failures and tragedies of the commons by *prohibiting* or *taxing* some types of economic activities, because they are contrary to widespread public interests such as mitigating climate change and/or adapting to its effects. Because many solutions are dependent on the development and adoption of alternative technologies, subsidies for technological innovation programs can also be developed. Familiar examples for automobiles are fuel efficiency regulations, taxes on gasoline and diesel fuel, and subsidies for electric vehicles.

Thus, economic, political and technological factors, along with information from climate science, all interact in decision-making about market failures and tragedies of the commons. The specific patterns and consequences of such interactions vary across economic sectors and industries as well as political systems. Yet, these interactions pose centrally important issues that pervade discourse and decision-making about what should be done about climate change.

12.7 Conclusion

There are recurrent themes in this concluding chapter and the preceding chapters as well; some are more explicit than others, and some more documented by empirical analysis. However, altogether, the analyses from many relevant disciplines in the wide-ranging topical focuses of the chapters lead to the following summary conclusions:

- The evidence from climate science is that a multitude of climate change *problems* are already present and are projected to be worse for at least the rest of the current decade. The frequency and severity of climate-related extreme weather events have been increasing for many years. Global temperatures as well as concentration levels of greenhouse gases have been increasing, and they are projected to continue to increase. Unless there are major reductions in fossil fuel emissions during the present decade, tipping points will likely occur, with catastrophic consequences.
- Many *solutions* are already in the process of being implemented—as a result of the leadership of some government, business, and civil society organizations. The solutions include a complex combination of technological, economic, and political changes. In some instances, increasing understanding of the problems and solutions—and changes in ethically based perceptions—have led to major policy changes by some governments and businesses. However, just as there are *leaders*, there are also *laggards* among governmental, business and other organizations.
- As for the future, the key questions are whether the *speed and scale of mitigation actions* to reduce emissions levels and *adaption actions* to reduce the impacts of past and continuing emissions will be sufficient to avoid the catastrophic projections of future climate change developments. There will undoubtedly be a combination of responsible and resourceful actions from many governments, businesses and societies—and also the opposite from others. In some societies, inter-generational differences in knowledge of the problems and support for the solutions will overcome the deniers, doubters and delayers?
- Inter-disciplinary analytic and policy approaches to the problems and solutions are needed. The issues are so complex that inter-disciplinary thinking is needed. Whether it is sufficient will be determined by the actions taken by governments, corporations and other non-governmental organizations, as well as populations and individual leaders.

Annex 12.1: Climate Change Models: Features, Issues and Applications

For several decades, much of the research by climate scientists has been based on computerized mathematical modelling analyses that may seem arcane to many readers. The purpose of this annex is to make the topic more accessible by drawing upon discussions from diverse sources, including Carbon Brief (2018), Climate Science (2022), Engels et al. (2023), Eltahir and Krol (2021), EU Copernicus (2023a, 2023b), Harper (2018), IPCC (2021c), MIT (2023) and National Centre for Atmospheric Science (2023).

Components/elements
Climate change models can include one or more of the "major components of the climate system: atmosphere, ocean (including sea level change), land, biosphere and cryosphere, and the carbon, energy and water cycles" (IPCC, 2021b). These components—which are sometimes referred to as "elements"—can be represented by many different indicators of climate conditions and changes in them.

Uses
Modeling exercises can be used to (a) increase understanding of present climate conditions by constructing and testing models based on evidence about past conditions, or (b) project future climate conditions based on past and current climate conditions. It has been noted that a "projection" is "a statement of probability that says something could happen in the future *if* certain conditions develop," while a "prediction" is "a statement of probability that says something will happen" (Climate Science, 2022).

Variables
The kinds of variables in climate models include atmospheric clouds, aerosols, and gasses; land surface snow and ice, lakes and rivers, soil, and vegetation; oceans and sea ice. A broader kind of model—an "earth systems" model—includes many climate-related variables but also others such as "landscape use" that can be useful in understanding the consequences of climate change (US DOE, 2023). There is interest in expanding the "physical" variables in climate models to include "social" variables, such as climate protests and social movements, transnational initiatives, corporate responses, government regulations, and consumption patterns. Among them, studies have found that corporate responses and consumption patterns "continue to undermine the pathways to decarbonization" (Engels et al. 2023).

Precision
Precision issues arise in relation to space and time. What precisely are the locational and temporal dimensions of the source data and the findings? The size of the three-dimensional "grid" indicates the distances between observation points. Precision in that sense has increased significantly in the early 2020s.

Uncertainty: likelihood and level of confidence

There are inevitably issues about the uncertainty of models. In IPCC discussions of models, the terms "likelihood" and "levels of confidence" are routinely and systematically applied to address these issues (IPCC, 2021a: 4). Accordingly, "The likelihood of an outcome or result" has the following probabilistic equivalents:

"virtually certain" equals 99–100% probability
"very likely" equals 90–100%
"likely" equals 66–100%
"about as likely as not" equals 33–66%
"unlikely" equals 0–33%
"very unlikely" equals 0–10%
"exceptionally unlikely" equals 0–1%

The following terms, which are consistent with the above, are also sometimes used:

"extremely likely" equals 95–100%
"more likely than not" equals > 50–100%
"extremely unlikely" equals 0–5%

A "level of confidence" in a finding, according to IPCC convention, is expressed using five qualifiers: "very low, low, medium, high and very high…." (Kause et al. 2022). The level of confidence is thus partly based on the judgments of authors of IPCC reports, whereas the "likelihood" indicated above is derived on the basis of inferential statistics that focus on the probabilities of sample findings.

Updating data and models

As an on going scientific enterprise, the development of climate models continues along many dimensions. The collection of additional data from recent and current climate-related events keeps data sets up-to date; this data comes from satellites, on-ground monitoring systems, and oceanic monitoring systems. These and other sources of data can be added to specialized data sets of individual forcing agents. For instance, there has been growing data collection concerning methane emissions from abandoned coal mines and oil wells. There has also been increased data gathering activity in both the Arctic and Antarctic regions because of expanding concern about the effects the melting ice on sea level rise. In the Arctic, deposits of black carbon particles on the glaciers in Greenland are especially significant (Brewer, 2019; Council, 2021). Data sets can also be augmented by increasing measurements of past temperatures and other climate conditions—extending back thousands of years and beyond. Data from tree rings, ice cores, and sea bottoms can be used in these data collection processes. "Earth System" models include inter-acting contextual variables in addition to the variables included in climate change models, as described above (Heavens et al. 2013).

Example from Copernicus

An example of the applications of climate change models to ocean temperatures, sea levels and other features is the "The Operational Mercator global ocean analysis and forecast system" of Copernicus (2023a; used with permission; also see Copernicus (2023b); it provides temperature, sea level and other information, including:

> 10 days of 3D global ocean forecasts updated daily. The time series is aggregated in time in order to reach a two full years' time series sliding window.
>
> This product includes daily and monthly mean files of temperature, salinity, currents, sea level, mixed layer depth and ice parameters from the top to the bottom over the global ocean. It also includes hourly mean surface fields for sea level height, temperature and currents.
>
> The global ocean output files are displayed with a 1/12 degree horizontal resolution with regular longitude/latitude equirectangular projection.
>
> 50 vertical levels are ranging from 0 to 5500 meters.

Annex 12.2: Sector-Specific Policies and Priorities for Action

Energy. Reducing GHG emissions across the full energy sector requires major transitions, including a substantial reduction in overall fossil fuel use, the deployment of low-emission energy sources, switching to alternative energy carriers, and energy efficiency and conservation. The continued installation of unabated fossil fuel infrastructure will 'lock-in' GHG emissions.

> In this context, 'unabated fossil fuels' refers to fossil fuels produced and used without interventions that substantially reduce the amount of GHG emitted throughout the life cycle; for example, capturing 90% or more CO_2 from power plants, or 50–80% of fugitive methane emissions from energy supply.

Industry. Reducing industry emissions will entail coordinated action throughout value chains to promote all mitigation options, including demand management, energy and materials efficiency, circular material flows, as well as abatement technologies and transformational changes in production processes. Progressing towards net zero GHG emissions from industry will be enabled by the adoption of new production processes using low- and zero-GHG electricity, hydrogen, fuels, and carbon management.

Buildings. Existing buildings, if retrofitted, and buildings yet to be built, are projected to approach net zero GHG emissions in 2050 if policy packages, which combine ambitious sufficiency, efficiency, and renewable energy measures, are effectively implemented and barriers to decarbonisation are removed.

Transportation. Demand-side options and low-GHG emissions technologies can reduce *transport* sector emissions in developed countries and limit emissions growth in developing countries .... Demand-focused interventions can reduce demand for all transport services and support the shift to more energy efficient transport modes .... Electric vehicles powered by low-emissions electricity offer the largest decarbonisation potential for land-based transport, on a life cycle basis .... Sustainable biofuels can offer additional mitigation benefits in land-based transport in the short and medium term .... Sustainable biofuels, low-emissions hydrogen, and derivatives (including synthetic fuels) can support mitigation of CO_2 emissions from shipping, aviation, and heavy-duty land transport but require production process improvements and cost reductions. Many mitigation strategies in the transport sector would have various co-benefits, including air quality improvements, health benefits, equitable access to transportation services, reduced congestion, and reduced material demand.

Agriculture Forestry and Other Land Use (AFOLU). AFOLU mitigation options, when sustainably implemented, can deliver large-scale GHG emission reductions and enhanced removals, but cannot fully compensate for delayed action in other sectors. In addition, sustainably sourced agricultural and forest products can be used instead of more GHG-intensive products in other sectors. Barriers to implementation and trade-offs may result from the impacts of climate change, competing demands on land, conflicts with food security and livelihoods, the complexity of land ownership and management systems, and cultural aspects. There are many country-specific opportunities to provide co-benefits (such as biodiversity conservation, ecosystem services, and livelihoods) and avoid risks (for example, through adaptation to climate change).

Source: Excerpted and compiled by the author from IPCC (2022).

Annex 12.3: Systemic Shocks

Shocks from other "domains" sometimes alter the future paths of the scientifically based analyses. Understanding history is thus relevant, especially when there are "shocks" of various types that fundamentally alter important patterns, trends and projections. There are many examples from the past few decades of climate change, and there will surely be more in coming years.

Engineers and economists are familiar with disruptive technologies that have significant economic consequences. In the context of climate change issues, for instance, the development of fracking technologies applied to drilling natural gas wells has changed the relative costs of natural gas and coal as fuels for electric power plants. The shock effects of disruptive technologies can thus create both winners and losers within industries. The changes in the relative prices of natural gas and coal, however, have not only been economically significant for those industries and the areas where they are the predominant industries; they have also had significant implications for carbon dioxide and other emissions.

Economic shocks in the form of the ups and downs of business cycles also have impacts on climate change. In the up phase, consumption of electricity, motor vehicle fuels and other carbon intensive activities increase; in the down phase, of course, the opposite occurs. The relationship, however, can be complicated by the confounding influence of other factors.

Political shocks with climate change consequences are also familiar—some of them domestic political shifts, others international events. For instance, the 2016 election in the US resulted in dramatic changes in climate change policies from the Obama to the Trump presidencies, as did the 2020 election as a result of which the Biden administration reversed a broad array of the previous administration's policies (Brewer, 2022).

A *public health* shock in the form of the covid pandemic of 2021–2022 illustrates yet another example of how shocks from outside the normal scope of climate change issues can have significant effects on climate change—and interact with other shocks to the climate change system. In particular, the trend of increasing global carbon dioxide emissions was reversed briefly during 2020–2021, as the peak of the pandemic caused a global economic decline, including in air travel. Then, as the global economy began to recover in 2022, including resumption of previous and increasing levels of air travel, global emissions resumed their previous levels and trends from before the covid shock.

The Russian-Ukrainian war represents another kind of systemic shock—one from an action from the contemporary international relations system. There are two sets of issues: one concerns specifically the effects on the energy sector; the other concerns the implications for diplomatic relations generally, including climate change issues. As for the energy sector, the immediate effects included increased prices for oil and natural gas, with the implications for consumption levels of those and alternative fuels to produce electricity. The longer-term issues concerned the effects on nuclear power as a source of electricity. There were also efforts to reassert and realign international cooperation on climate issues in the context of the disruption to existing efforts resulting from the war in Europe. (See Chap. 3 for information about the impact on energy and climate change policies in Europe.)

Applications of Artificial Intelligence (AI) to climate change issues exemplify yet another kind of shock, namely an analytic shock. As AI has become a contentious topic in public political discourse in the early 2020s, its application to climate change topics has also become more frequent and controversial, especially in Europe and North America. The potential contributions and risks of AI are especially relevant in the context of this chapter's focus on the future. The urgent need for climate change action together with the emergence of a new IT analytic toolkit make the intersection of the two topics a particularly apt topic of this concluding chapter. As of 2023, these topics were especially salient in Europe; there was also growing concern in the US about many questions about AI. A rapidly expanding research agenda was evident at universities and think tanks in many countries, as the references list indicates. One result has been a proliferation of articles, presentations at professional conferences, and new websites that feature blogs and reports

Table 12.5 Comparative estimates of years until carbon budget depletion

Source	Years until carbon budget depletion[a]
ICCT	30
Liu et al. (2023): Adjustment for projected emission rate	25.8
Liu et al. (2023): Adjustment for projected emission rate and base year	23.7

Sources Excerpted and compiled by the author from (Liu et al. 2023), which reports data from IPCC (2021c)

[a]Assumes Paris 2.0 or less temperature increase target. Also see the article for other targets and base year data

by AI experts (Economist, 2023; Kaack et al. 2022; OECD, 2023; Rolnick, 2022; Rolnick et al. 2019). At the same time, however, in the larger context of controversies about AI, there have been apocalyptic warnings from many AI experts that other kinds of consequences are possible (Center for AI Safety, 2023; Valance, 2023). Assessments of potential contributions of AI to analyses of climate change issues will need to take into account the evolution of the broader issues about the risks of AI.

Annex 12.4: Data About the Earth's Carbon Budget

Research including comparative full year 2022 global CO_2 emissions data in a time series indicates the need to shorten estimates of the earth's carbon depletion dates (Liu et al. 2023). Annex Table 12.5 compares the revised estimates with those from the ICCT. The revisions thus include 2022 data that were not previously available. The revisions include both higher levels of CO_2 already accumulated in the budget and higher levels of projected annual emissions in the future. The table only includes a few key data points. Readers are encouraged to consult the original article for additional information.

It is important also to emphasize that "When considering the non-CO_2 contributors to anthropogenic warming, such as methane, nitrous oxide, and fluorinated gases, the remaining carbon budget becomes even smaller" (Liu et al. 2023). One could add black carbon particulate matter to the list, since it is one of the three biggest contributors to global warming along with carbon dioxide and methane.

Annex 12.5: The Problematic Paradox of Atmospheric Cooling Above the Troposphere

"There is a paradox at the heart of our changing climate. While the blanket of air close to the Earth's surface is warming, most of the atmosphere above is becoming dramatically colder" (Pearce, 2023; also see Santer et al. 2023). This paradox has been noted for many years by climate scientists, and recent data collected by satellites has confirmed the paradox (Mlynczak, 2021).

It is important to recognize two basic facts about the paradox: one is that CO_2 emissions are a cause, and the other is that the temperature changes have been large—about 2 °C (or about 3 °F) between 2002 and 2019 and as much as 7–8 °C (13–14 °F) during the present century. There are significant implications for climate change of these trends. In particular, instabilities in the stratosphere can exacerbate extreme weather events in the northern hemisphere caused by the polar jet stream, such as "persistent intense rains to summer droughts and 'blocking highs' that can cause weeks of intense cold winter weather from eastern North America to Europe and parts of Asia" (Pearce, 2023).

Another effect is that the ozone hole over the Arctic is likely to increase (Von der Gathen et al. 2023). Unlike the diminishing ozone hole over Antarctica, the growing Arctic hole exposes large populations in the northern hemisphere to the increased health threat. This is thus another example of the confluence of climate and health issues—and therefore the value of inter-disciplinary analytic approaches and the need for policymakers to be aware of the potential co-benefits of many climate actions.

Questions to Ponder

1. What do you think are the strengths and weaknesses of the models? Why?
2. Does any particular model seem the best one to you? Why?
3. Governments, businesses and civil society have different roles in various potential future paths to address climate change? What are the *distinctive* features of each?
4. One way to imagine the future is to ponder whether the world will "wise up" about climate change. What do you think?
5. Which technologies do you think are the most promising for reducing emissions in the future? Why?
6. Do you agree or disagree that the following major changes in the next few years must be achieved in order to avoid the most catastrophic impacts of climate change? Emissions from each of the following must be significantly reduced to net zero.
   Coal-fired power plants
   Diesel and gasoline motor vehicles
   Methane emissions in the energy sector and agriculture sector
   Black carbon emissions in the electric power and transportation sectors
7. Update any data in this chapter. What changes—if any—do you find in the patterns and/or trends?
8. There may be significant surprises about climate change problems and/or solutions in the future. Can you imagine any that would be different from the projections in this chapter? Why?

References

Alley, R., et al. (2003). Abrupt climate change. *Science, 299*, 5615. https://doi.org/10.1126/science.1081056. Accessed May 13, 2023.

American Geophysical Union. (2023). *The Greenland Ice Sheet is close to a melting point of no return, says new study.* https://phys.org/news/2023-03-greenland-ice-sheet.html. Accessed April 5, 2023.

Arctic Council. (2021). *Arctic states on track to reach the collective goal on black carbon emissions.* https://arctic-council.org/news/arctic-states-on-track-to-reach-the-collective-goal-on-black-carbon-emissions/. Accessed April 11, 2023.

Armstrong McKay, D., Staal, A., Abrams, J. F., Winkelmann, R., Sakschewski, B., Loriani, S., Fetzer, I., Cornell, S. E., Rockström, J., & Lenton, T. M. (2022). Exceeding 1.5°C global warming could trigger multiple climate tipping points. *Science, 377*, 6611. https://doi.org/10.1126/science.abn7950

Bator, F. (1958). The anatomy of market failure. *The Quarterly Journal of Economics, 72*(3), 351–379. https://doi.org/10.2307/1882231

Brewer, T. (2019). The Arctic Ocean's melting ice: Institutions and policies to manage black carbon. In P. G. Harris (Ed.), *Climate change and ocean governance* (Chap. 15). Cambridge University Press.

Brewer, T. (2022). *Transforming US climate change policies: 2021 and beyond.* Policy Brief. Springer. https://link.springer.com/book/9783030997151

Carbon Brief. (2018). *How do climate models work?* https://www.carbonbrief.org/qa-how-do-climate-models-work/. Accessed February 15, 2023.

Center for AI Safety. (2023). *Statement on AI risk.* https://www.safe.ai/statement-on-ai-risk#open-letter. Accessed May 30, 2023.

Climate Science. (2022). *Climate models: How can we project climate change?* https://climatescience.org/advanced-climate-climate-models. Accessed February 15, 2023.

Dangendorf, S., Marcos, M., Müller, A., Zorita, E., Riva, R., Berk, K., & Jensen, J. (2023). *Nature Communications.* https://www.nature.com/articles/s41467-023-37649-9. Accessed April 11, 2023.

Economist. (2023). Large, creative AI models will transform lives and labour markets. *The Economist.* https://www.economist.com/interactive/science-and-technology/2023/04/22/large-creative-ai-models-will-transform-how-we-live-and-work. Accessed May 7, 2023.

Eltahir, E., & Krol, A. (2021). *MIT climate portal. Climate models.* https://climate.mit.edu/explainers/climate-models. Accessed February 15, 2023.

Engels, A. et al. (2023). *Hamburg futures outlook 2023.* Cluster of Excellence Climate, Climate Change, and Society (CLICCS). https://www.cliccs.uni-hamburg.de/results/hamburg-climate-futures-outlook.html. Accessed March 3, 2023.

EU Copernicus. (2023a). *Copernicus marine service. Global ocean 1/12° physics analysis andforecast, updated daily.* https://data.marine.copernicus.eu/product/GLOBAL_ANALYSISFORECAST_PHY_001_024/notificSystemications. Accessed February 28, 2023.

EU Copernicus. (2023b). *Copernicus marine service. Global Ocean physics reanalysis.* https://data.marine.copernicus.eu/product/GLOBAL_MULTIYEAR_PHY_001_030/description. Accessed February 28, 2023.

Global Ocean Observing System. (2022). *ETOOFS expert team on operational ocean forecasting systems. Implementing operational ocean monitoring and forecasting systems: Summary booklet.* file:///D:/CH%2015%20NEW/ETOOFS%20Guide%20Summary%202022.pdf. Accessed February 27, 2023.

Guterres, A. (2022). *Secretary-general's remarks to high-level opening of COP27.* https://media.un.org/en/asset/k12/k12p31t118. Accessed December 12, 2022.

Hardin, G. (1968). The tragedy of the commons. *Science, 162*(3859), 1243–1248. https://doi.org/10.1126/science.162.3859.1243

Harper, S. L. (2018). *What are climate models and how accurate are they?* https://news.climate.columbia.edu/2018/05/18/climate-models-accuracy/. Accessed February 15, 2023.

Holdren, J. (2009). In A. Danner (Ed.), *On ABC news 'earth 2100': The final century of civilization?* https://abcnews.go.com/Technology/Earth2100/story?id=7697237&page=1. Accessed January 18, 2022.

Heavens, N. G., Ward, D. S., & Natalie, M. M. (2013) Studying and projecting climate change with earth system models. *Nature Education Knowledge , 4*(5), 4. https://www.nature.com/scitable/knowledge/library/studying-and-projecting-climate-change-with-earth-103087065/. Accessed April 11, 2023.

Höning, D., et al. (2023). Multistability and transient response of the Greenland ice sheet to anthropogenic CO_2 emissions. *Geophysical Research Letters.* https://phys.org/news/2023-03-greenland-ice-sheet.html. Accessed April 5, 2023.

IPCC. (2018). *Special report on global warming of 1.5°C.* https://www.ipcc.ch/sr15. Accessed March 16, 2023.

IPCC. (2021a). Summary for Policymakers. In *Climate change 2021: The physical science basis.* Contribution of Working Group I to the Sixth Assessment Report of the Intergovernmental Panel on Climate Change. https://www.ipcc.ch/report/ar6/wg1/downloads/report/IPCC_AR6_WGI_SPM.pdf. Accessed March 3, 2023.

IPCC. (2021b). Technical summary. In *Climate change 2021: The physical science basis.* Contribution of Working Group I to the Sixth Assessment Report of the Intergovernmental Panel on Climate Change. https://www.ipcc.ch/report/ar6/wg1/downloads/report/IPCC_AR6_WGI_TS.pdf. Accessed February 27, 2023.

IPCC. (2021c). *Climate change 2021: The physical science basis.* Contribution of Working Group I to the Sixth Assessment Report of the Intergovernmental Panel on Climate Change. https://www.ipcc.ch/report/ar6/wg1/. Accessed February 27, 2023.

IPCC. (2022). Summary for policymakers. In P. R. Shukla, J. Skea, R. Slade, A. Al Khourdajie, R. van Diemen, D. McCollum, M. Pathak, S. Some, P. Vyas, R. Fradera, M. Belkacemi, A. Hasija, G. Lisboa, S. Luz, J. Malley, (Eds.), *Climate change 2022: Mitigation of climate change.* Contribution of Working Group III to the Sixth Assessment Report of the Intergovernmental Panel on Climate Change. Cambridge University Press. https://doi.org/10.1017/9781009157926.001

IPCC. (2023). *Synthesis report of the IPCC sixth assessment report (AR6), summary for policymakers.* https://report.ipcc.ch/ar6syr/pdf/IPCC_AR6_SYR_SPM.pdf. Accessed March 21, 2023.

Kaack, L. H., et al. (2022). Aligning artificial intelligence with climate change mitigation. *Nature Climate Change,* 12, 518–527. https://www.nature.com/articles/s41558-022-01377-7. Accessed May 7, 2023.

Kause, A., et al. (2022). Confidence levels and likelihood terms in IPCC reports: a survey of experts from different scientific disciplines. *Climatic Change, 173*, 2. https://doi.org/10.1007/s10584-022-03382-3

Kemp, L., et al. (2022). Climate endgame: Exploring catastrophic climate change scenarios. *PNAS, 119*, 34. https://doi.org/10.1073/pnas.2108146119

Keohane, N., & Olmstead, S. (2016). *Markets and the environment* (2nd ed.). Island Press.

Klower, M., et al. (2021). Quantifying aviation's contribution to global warming. *Environmental Research Letters, 16*, 10. https://doi.org/10.1088/1748-9326/ac286e

Lenton, T. M. (2019). Climate tipping points—Too risky to bet against. *Nature, 575*, 592–595. https://www.nature.com/articles/d41586-019-03595-0. Accessed April 23, 2023.

Li, Q., England, M. H., Hogg, A. M., et al. (2023). Abyssal ocean overturning slowdown and warming driven by meltwater. *Nature, 615*, 841–847. https://www.nature.com/articles/s41586-023-05762-w. Accessed April 23, 2023.

Lindwall, E. (2022). *Rich, polluting nations still owe the developing world.* https://www.nrdc.org/stories/how-ditch-biggest-fossil-fuel-offenders-your-life. Accessed May 13, 2023.

Liu, Z., et al. (2023). Monitoring global carbon emissions in 2022. *Nature Reviews, 4*, 205–206. https://www.nature.com/articles/s43017-023-00406-z

MIT. (2023). *Earth system model*. https://globalchange.mit.edu/research/research-tools/earth-system-model. Accessed February 15, 2023.

Mlynczak, M. G. (2021). An observational gap at the edge of space. *Eros*. https://eos.org/opinions/an-observational-gap-at-the-edge-of-space. Accessed May 20, 2023.

NASA and CNES. (2023). *NASA uses 30-year satellite record to track and project rising seas*. https://climate.nasa.gov/news/3257/nasa-uses-30-year-satellite-record-to-track-and-project-rising-seas/. Accessed April 6, 2023.

National Centre for Atmospheric Science (NCAS). (2023). *What is a climate model?* https://ncas.ac.uk/learn/what-is-a-climate-model/. Accessed February 21, 2023.

National Snow and Ice Data Center (NSIDC). (2023a). *Near-real-time DMSP SSMIS daily polar gridded sea ice concentrations, version 2 (NSIDC-0081)*. https://nsidc.org/data/nsidc-0081/versions/2. Accessed February 28, 2023.

National Snow and Ice Data Center (NSIDC). (2023b). *Arctic [and Antarctic] Sea ice news & analysis*. https://nsidc.org/arcticseaicenews/. Accessed April 21. 2023.

OECD. (2022). *Climate tipping points: Insights for effective policy action*. https://www.oecd.org/environment/climate-tipping-points-abc5a69e-en.htm. Accessed March 16, 2023.

OECD. (2023). *OECD AI principles overview*. https://oecd.ai/en/ai-principles. Accessed May 7, 2023.

Ostrom, E. (1990). *Governing the commons: The evolution of institutions for collective action*. Cambridge University Press.

Pearce, F. (2023). *The upper atmosphere is cooling, prompting new climate concerns*. Yale Environment 360. https://e360.yale.edu/features/climate-change-upper-atmosphere-cooling. Accessed May 20, 2023.

Rolnick, D. (2022). *The paper that launched CCAI*. https://www.climatechange.ai/blog/2022-03-08-tccmi-publication. Accessed April 10, 2023.

Rolnick, D., Donti, P., et al. (2019). Tackling climate change with machine learning. *Computer Science*. https://arxiv.org/abs/1906.05433. Accessed April 10, 2023.

Salanié, B. (2000). *The microeconomics of market failures*. MIT Press.

Santer, B., et al. (2023). Exceptional stratospheric contribution to human fingerprints on atmospheric temperature. *Proceedings of the National Academy of Sciences, 120*(20), e2300758120. https://doi.org/10.1073/pnas.2300758120

Sinet, S., von der Heydt, A. S., & Dijkstra, H. A. (2023). AMOC stabilization under the interaction with tipping polar ice sheets. *Geophysical Research Letters, 50*, 2. https://doi.org/10.1029/2022GL100305

Songwe, V., Stern, N., & Bhattacharya, A. (2022). *Finance for climate action: Scaling up investment for climate and development*. Grantham Research Institute on Climate Change and the 12 Environment, London School of Economics and Political Science. https://www.lse.ac.uk/granthaminstitute/wp-content/uploads/2022/11/IHLEG-Financefor-Climate-Action.pdf. Accessed January 10, 2023.

Stavins, R. (2001). *Environmental economics and public policy*. Edward Elgar.

Stavins, R. (2013*) Economics of climate change and environmental policy*. Edward Elgar.

Stavins, R. (2019). *Economics of the environment* (7th ed.). Edward Elgar.

Stern, N. (2006). *The economics of climate change: The Stern review*. https://www.lse.ac.uk/GranthamInstitute/publication/the-economics-of-climate-change-the-stern-review/. Accessed May 22, 2023.

Stern, N., & Romani, M. (2023). *The global growth story of the 21st century: Driven by investment and innovation in green technologies and artificial intelligence*. Grantham Research Institute on Climate Change and the Environment, London School of Economics and Political Science and Systemiq.

Stiglitz, J. (1989). Markets, market failures, and development. *The American Economic Review, 79*(2), 197–120. Accessed December 28, 2022.

Systemiq. (2021). *The Paris effect—COP26 edition*. https://www.systemiq.earth/wpcontent/uploads/2021/11/The-Paris-Effect-COP26-edition-SYSTEMIQ.pdf

US NOAA. https://www.gfdl.noaa.gov/climate-modeling/. Accessed February 15, 2023.

US DOE. (2023). *DOE explains … Earth system and climate models*. https://www.energy.gov/science/doe-explainsearth-system-and-climate-models. Accessed February 15, 2023.

Vallance, C. (2023). *Artificial intelligence could lead to extinction, experts warn*. BBC News. https://www.bbc.com/news/uk-65746524. Accessed May 30, 2023.

Von der Gathen, P., et al. (2023). Climate change favours large seasonal loss of Arctic ozone. *Nature Communications, 12*(1). https://doi.org/10.1038/s41467-021-24089-6

Yin, J. (2023). Rapid decadal acceleration of sea level rise along the U.S. East and Gulf Coasts during 2010–2022 and Its Impact on Hurricane-Induced Storm Surge. *Journal of Climate*, 1–38. https://doi.org/10.1175/JCLI-D-22-0670.1

Keeping up with Future Developments

Copernicus. *Open access hub*. https://scihub.copernicus.eu/dhus/#/home

NASA. *Ask NASA climate*. https://climate.nasa.gov/explore/ask-nasa-climate/?page=0&per_page=40&order=publish_date+desc%2C+created_at+desc&search=&hide_filter_bar=true&grid_list_klass=full_news_list&category=25

International Organizations

IEA. *Climate change*. https://www.iea.org/topics/climate-change

Intergovernmental Panel on Climate Change (IPCC). https://www.ipcc.ch/about/

International Institute for Applied Systems Analysis (IIASA). https://iiasa.ac.at/

IRENA. *Climate action*. https://www.irena.org/Energy-Transition/Country-engagement/Climate-Action

OECD. *Climate change*. https://www.oecd.org/climate-change/

UNEP. *About climate change*. https://www.unep.org/explore-topics/climate-change/about-climate-change

UNCTAD. *Climate change*. https://unctad.org/topic/trade-and-environment/climate-change

UNFCCC. https://unfccc.int/essential-information

WMO. *Climate*. https://public.wmo.int/en/our-mandate/climate

World Bank. *Climate knowledge portal*. https://climateknowledgeportal.worldbank.org/

Also see the reference lists at the ends of relevant chapters.

Refereed Journals

Climate. https://www.mdpi.com/journal/climate

Climatic Change. https://www.springer.com/journal/10584

Climate Policy. https://www.tandfonline.com/journals/tcpo20

Journal of Climate. https://www.ametsoc.org/ams/index.cfm/publications/journals/journal-of-climate/

Nature Climate Change. https://www.nature.com/nclimate/journal-information

Also see the reference lists at the ends of relevant chapters.

Printed in the USA
CPSIA information can be obtained
at www.ICGtesting.com
LVHW050727130224
771713LV00003B/158

9 783031 429057